Impact of Marine Pollution on Society

VIRGINIA K. TIPPIE
DANA R. KESTER
Editors

Center for Ocean Management Studies

University of Rhode Island

PRAEGER SPECIAL STUDIES • PRAEGER SCIENTIFIC
A J.F. BERGIN PUBLISHERS BOOK

Library of Congress Cataloging in Publication Data
Main entry under title:
Impact of marine pollution on society.
  "Papers presented at the fourth annual
conference organized by the Center for Ocean
Management Studies, University of Rhode Island"
—Introd. Bibliography: p.
  1. Marine pollution—Social aspects—Congresses.
I. University of Rhode Island.
Center for Ocean Management Studies.
GC1081.I48   363.7'394   81-12270
ISBN 0-03-059732-3   AACR2

Copyright © 1982 by J. F. Bergin Publishers, Inc.

J. F. Bergin Publishers, Inc.
670 Amherst Road
South Hadley, Massachusetts 01075

Published in 1982 by Praeger Publishers
CBS Educational and Professional Publishing
A Division of CBS, Inc.
521 Fifth Avenue, New York, New York 10175, USA

0123456789   056   987654321

Printed in the United States of America

Contents

Preface

There has been considerable research on marine pollution; however, few studies have addressed its impact on society. As we recognize that marine "pollution" implies a degradation of environmental quality as perceived by society, it becomes necessary to assess its impact from scientific, economic, and policy perspectives. To provide an opportunity to share insights on this important interdisciplinary subject, the Center for Ocean Management Studies at the University of Rhode Island convened a conference on the *Impact of Marine Pollution on Society*. This volume is a compilation of the invited papers and conference discussions.

The units and chapters in the book generally reflect the conference program. Each unit is comprised of three invited papers followed by the commentary of a provocateur and a general discussion. Unit One, "Status of Marine Pollution," outlines the political and social framework for controlling pollution, the development of our understanding of the human impact on the marine environment, and international efforts to assess and control marine pollution. Subsequent units examine three case studies of accidental, "intentional," and chronic pollution from scientific, economic, and policy perspectives. The case studies include the *Amoco Cadiz* oil spill on the northwest coast of France, the disposal of contaminated dredged materials in Long Island Sound, and the chronic discharge of sewage sludge and effluent off the coast of southern California. In the last unit, "Future Prospects and Strategies," the authors discuss the need to link scientific research to the concept of public interest, and the importance of translating our understanding of marine pollution into human terms or values. Although it is recognized that the ocean has a great ability to assimilate many wastes, the challenge that lies ahead is to assure both effective utilization and protection of the coastal and marine environment.

The conference was sponsored by the Center for Ocean Management Studies at the University of Rhode Island in cooperation with the National Marine Pollution Program Office. The Center for Ocean Management Studies was created in the fall of 1976 for the purpose of promoting effective coastal and ocean management. This is achieved by providing an interdisciplinary forum for communication, research, and education on marine issues. We feel that this volume represents such a forum. By better understanding the interdisciplinary aspects of marine pollution, hopefully, we can more effectively manage our marine environment.

Looking to the future, it will become increasingly important to distinguish between the *disposal of wastes* in the ocean and *pollution of the ocean*. The input of some wastes may not result in a degradation of the marine environment; and, even more significantly, it is likely that some wastes may be placed in the ocean with less severe consequences to the environment than if they were to enter the fresh-water hydrological cycle or the atmosphere. A primary challenge for marine scientists, economists, and managers, therefore, is to determine the extent to which the ocean can recycle specific wastes, returning them to the biogeochemical, which has regulated the composition of the oceans for millions of years.

*   *   *

Many people contributed to this volume. First and foremost, we would like to thank the members of the program committee who served as the session chairmen: Thomas Grigalunas, Resource Economics, University of Rhode Island; John Knauss, Provost for Marine Affairs, University of Rhode Island; Saul Saila, Oceanography, University of Rhode Island; and Eric Schneider, Center for Ocean Management Studies, University of Rhode Island. In addition, a special thanks should be given to the conference speakers, who made this volume possible.

Numerous individuals and organizations provided logistical and technical support in organizing the conference and preparing the proceedings. However, special recognition should go to Carol Dryfoos, COMS Administrative Assistant; Nancy Ingham, COMS Technical Editor; and the University of Rhode Island Conference Office. Last, but not least, special thanks is extended to the National Marine Pollution Program Office, National Oceanic and Atmospheric Administration, for funding the conference.

Virginia K. Tippie
Executive Director
Center for Ocean Management Studies

Dana R. Kester
Chairman
Conference, "The Impact of Marine Pollution on Society"

UNIT ONE

Status of Marine Pollution

WHAT EVIDENCE IS THERE THAT MARINE POLLUTION HAS AN IMPACT ON SOCIETY? In considering this question, one should distinguish between the public perception and the technical assessments of marine pollution. We have seen shifts in fisheries populations, and in some cases the public perception has been that these changes are linked to marine pollution. In many cases this cause and effect relationship is not justified on technical grounds because we have insufficient understanding of the natural variability of fish populations.

Another example in which public perception has attributed an environmental event to marine pollution is the 1976 anoxia off the coast of New Jersey and the subsequent fish kill. Some people concluded that this event resulted from the ocean dumping of wastes in the New York Bight. Technical assessments of this anoxia, however, indicate that it could be accounted for by natural processes.

These two examples emphasize the need to understand not only what pollutants do to the environment but also the natural processes occurring in the environment in order to assess the impact of pollution on man and the environment.

There are two cases in which we have clear indications that marine pollution has had an impact on society. One of these is in the closure of shell fisheries grounds because of bacterial contamination from sewage wastes in a number of areas of the country. This is an impact which has been gradual and surprisingly accepted by society in much the same way as air pollution has been gradual, accepted and therefore tolerated to a fairly high degree. It would be useful to know the costs of this effect on the environment and on man and to compare these with the costs to eliminate this pollution.

A second case in which we can demonstrate the impact of marine pollution on society is in the behavior of synthetic organic chemicals, which have tended to accumulate in sediments of shallow waters. The occurrence of PCBs in sediments of the Hudson River and of the New Bedford Harbor, Massachusetts, and of Kepone in the James River, Virginia, have subsequently affected our use of these environments. Thus, there is significant evidence that marine pollution affects not only the environment but also, to a significant degree, man's further utilization of the marine environment.

In this book we attempt to examine the evidence more thoroughly and set a sense of direction for the future management of activities in the marine environment.

Dana R. Kester
University of Rhode Island

CHAPTER 1

The Political Context

ANNMARIE H. WALSH
Institute of Public Administration
United States

This nation's endeavor to achieve clean waters, which began in earnest
in 1972, is in danger of repeating a familiar pattern: a highly publicized,
enthusiastic, legislative beginning with grand goals, followed by rapid
dissipation of political energies, weak implementation of programs, inadequate evaluation, public disillusionment, and premature relegation
of the issue to a low political priority.

Unrealistic goals tend to produce political failure. While many of
the weaknesses of programs to reduce marine pollution can be attributed
to persistent characteristics of the U.S. government and politics, the
scientific community bears a special burden in this case. Definition of
achievable goals and of the expected costs and benefits of reaching
them; the generation of the data needed to make these assessments;
and the translation of analysis into terms comprehensible to the public
and usable by policy-makers and implementers are the three challenges
that must be fulfilled by the experts on marine environments if political
energy is to be effectively channeled toward controlling pollution.

The history of public policy on marine pollution is reminiscent of
the history of manpower policy. In 1946, the Congress of the United
States officially declared a national policy of full employment. Agencies
and programs have been created, reorganized, reoriented and renamed
many times in the ensuing thirty-four years. Not only are we no closer
to the goal, we have never defined intermediary targets or committed
the public resources necessary to their achievement.

In 1972, Congress declared that our navigable and ocean waters
should be of a "quality which provides for the protection and propagation of fish, shellfish and wildlife, and provides for the recreation in
and on the water" by 1983. By 1985, all discharge of pollutants into
navigable waters was to be eliminated. Congressional enthusiasm for
clean waters was sufficient to create willingness to override presidential
veto of amendments to the federal Water Pollution Control Act.[1] However, the portent of continuing obstacles was the impoundment of funds

by the Executive. Subsequent events freed logjammed monies but do not inspire confidence that the obstacles have been surmounted.

In early 1979, the President's chief advisor on inflation and the chairman of the Council of Economic Advisors issued a memorandum which concluded that the existing water quality rules were "prohibitively expensive." It called for reexamination of them "to relate costs to benefits."[2] This perspective reflected a dramatic shift from a policy goal of clean waters by a certain date to supporting only such improvements for which benefits demonstrably outweigh costs. This shift could dramatically slow public action because no framework has been developed to identify comprehensively the costs and benefits of marine pollution control, let alone to calculate them in aggregate net terms. At the same time, the existing water quality rules have been characterized by the U.S. General Accounting Office and by environmental groups as totally inadequate. The federal agencies responsible for achieving clean waters are buffeted by pressures from opposite directions.

THE POLITICAL STATUS OF POLLUTION CONTROL

"Man marks the earth with ruin," proclaimed the poet Byron. A contemporary author has asked, "Are we perhaps fated to mark the ocean with ruin, to plunder, pollute and contend until we have a ghost ocean bereft of all but the voice of the waters? . . . To avoid such a fate will require comprehensive planning and policy making. Yet before national planning can take place, we must understand the ocean, its tolerances and stresses as well as its wealth."[3]

The progress toward understanding the ocean that is being made by the scientific community — described in such documents as the report on the *Assimilative Capacity of U.S. Coastal Waters for Pollutants*[4] — is not being communicated effectively to either the general public or the policy makers. Technical bibliographies are growing rapidly; public understanding is not.

Progress *has* been made however. Some $50 billion has been spent in a decade upon wastewater treatment facilities. Ocean dumping has been subjected to licensing processes; $800 million in federal funds was authorized for regional water quality planning from 1973 to 1980. Of this amount, however, only about one-third had been obligated by September 30, 1978.[5] Lack of funds is not the major obstacle to implementation.

Actual results obtained for the dollars authorized have been limited, and prognoses for the near future are discouraging. This can be illustrated by the cases of water quality planning and activities undertaken to meet the national deadline on ocean dumping in New York City.

The Status of Areawide Water Quality Planning:

Section 208 of the federal Water Pollution Control Act (amendments of 1972) provides for "comprehensive planning to be applied to water quality policy." In addition, Section 201 provides for planning with regard to site-specific facilities. But federal policy called for "comprehensive" water quality planning to be assigned to single-function local groups, which insures that it will *not* be comprehensive. Pollution in the New York Bight and economically feasible solutions to it are interstate phenomena. In the New York region there are several organizations with planning capacities including two interstate organizations: the Tri-State Regional Planning Commission, which has comprehensive planning staffs working in transportation, housing and economic development; and the Interstate Sanitation Commission, which conducts technical studies and monitoring of water pollution.

New York City was designated an areawide planning jurisdiction for water quality planning and embarked on its Section 208 program, with 100% federal funding, in 1976. The starting date lagged legislative authorization by four years and grant approval by one year. The terms of the legislation allowed two years for preparation of a plan which, among other tasks, would inventory both the physically specific and the general geographic sources of pollution. New York City submitted a draft report in April 1978, meeting the deadline only by omitting from the "plan" major aspects called for by federal guidelines. (In fact, 12 out of 16 of the Section 208 planning agencies surveyed by the General Accounting Office in 1978 had been unable to address most of the priority problems cited in Environmental Protection Agency (EPA) guidelines.)

New York City received 100 pages of comments from federal EPA officials and over 50 pages of comments from the state Department of Environmental Conservation. The plan was revised over the course of another year and subjected to public discussions and approval by the city's Board of Estimate in January 1980. The plan was held up, however, awaiting approval by the City Council, the city's chief legislative body. That approval is called for in order to assure financial commitment, within the terms of the city's financial plan, to implementation. The City Council took no action on the plan and the city risked loss of federal funds by failing to have an approved plan by June 30, 1980. The city negotiated with the federal officials and did receive some additional funds for water quality work, but other categories of funds for facilities planning have been held up since 1979 because of the lack of a certified plan.

The technical studies conducted for the city's Section 208 plan have proved useful to the specialists in waste management. They remain virtually unknown to the general public despite scores of public hearings, citizen advisory committee meetings and newsletters. Only threatened loss of funds raises the issue in the press, and that briefly.

The plan does not make an effective public case for the programs it recommends. The documents can be searched in vain for a clear exposition of what beach areas are closed or of limited use because of pollution, and of what population groups are damaged by this loss. The middle class has written off and abandoned huge tracts of waterfront real estate on the New York Bight, leaving them to marginal uses and slum populations with no political clout. The plan documents fail to catalogue this phenomenon.

The documents do not provide an explanation of impacts on the price and availability of fish and shellfish for food. They do not explain the impacts of occasional beach foulings on tourist and vacation economies, or the impacts of acid wastes and heavy metals on the food chain. To generate popular attention and political support, water quality management needs a counterpart to "cigarette smoking is dangerous to your health." Even if we do not have proof of specific cause and effect relationships, we must begin to document risks and losses in much more convincing ways if water quality management is to muster continued popular and political support.

In the meantime, the Section 208 plan is a political liability to New York City's officials. It attempts to earmark some long-range expenditure patterns, limiting the flexibility that the political officials seek in order to cope with continuing fiscal constraints and recurring fiscal "crises." Moreover, even the city's director of the Section 208 project believes the targets (now reinforced by court order) for building two major new treatment plants up to secondary treatment levels are unnecessarily high. Assuring the ability of fish to propagate — the official defined benefit of secondary-level treatment — is a low priority in a city with a shrinking budget (in adjusted dollars) and high crime and unemployment. Moreover, without resolution of how sludge produced by secondary treatment will be disposed of, the expected water quality improvements are unclear to the public. The two treatment projects are so far behind schedule that only the city's perpetually stalemated West Side Highway project elicits more discouragement. A decade of site planning with public participation has still not resolved conflicts over neighborhood dislocations associated with the large treatment plants. Finally, the city is painfully aware that no federal funds are available for operation and maintenance of facilities constructed. An extreme scenario contemplates local authorities failing to finance the operations of facilities constructed and attempting to force federal hands to distribute new subsidies.*

The citizens advisory committee to New York City's Section 208 Water Quality Planning program, for the most part enthusiastic supporters of clean waters, criticized the plan in convincing terms:[6]

1. The 208 plan fails to identify priorities and to rank targets.
2. It recommends a capital construction program beyond feasible financial capacity particularly when the incremental burdens of operating costs are factored in.
3. It fails to provide assessment of actual water quality improvements likely to result from recommended projects.
4. The plan does not explain or estimate the economic, aesthetic and recreational benefits that would be associated with revitalization of the city's vast waterfront.
5. The plan does too little to explore innovative techniques, such as those needed to eliminate raw sewage overspill from combined sewers during heavy rainfall.
6. The fragmentation of planning for regional waters, using separate agencies for New Jersey, Long Island, and major river bank areas, is not cost effective and discourages work that might lead to new methods rather than geographic application of known methods.

The Status of Ocean Dumping

A similar history applies to the issue of ocean dumping in the New York Bight. The volume of sludge being dumped continues to increase and the rate of increase to accelerate as the federal deadline for eliminating dumping approaches.[7] While the number of agencies dumping in the New York Bight has been reduced (from 250 municipal waste agencies in 1973 to less than 100, and from 150 industrial waste dumpers to 15), volume has increased by at least 50% since 1974 and is expected to triple by the year 2000.[8] Industrial volume is decreasing but it is unclear as to how much of that trend is due to reduction in manufacturing in the region rather than to improved waste management. Sludge loads still contain mercury and cadmium in excess of permissible limits. More-

---

*1977 amendments to the Clean Water Act do provide for an alternative approach for coastal communities. The burden of proof is put on them to show that disposal without "secondary treatment" will not harm ocean ecosystems. The manner of proof, however, is lacking from current methodologies.

over, Coast Guard surveillance of compliance with use of designated dump sites in the New York Bight is ineffective. A U.S. Comptroller General's study reported that the Coast Guard boarded no vessels and actually observed less than 1% of ocean dumping in 1975.[9] Coast Guard budget allocations give this function a low priority. The vast majority of apparent violations reported to EPA are not followed up.

Scientific testing of rates at which wastes can be absorbed by the ocean and related water quality monitoring have lagged. Some monitoring operations have been reduced due to budgetary constraints. Research indicating that several parameters of pollution (nitrogen, PCBs, and cadmium) are close to unacceptable limits in the New York Bight has not been translated into clear public criteria on content and volumes of sludge or dredge wastes.

In August 1980, the city held hearings on an "Accelerated Interim Sludge Management Plan." It was prepared because the city was required to show a strategy for meeting the December 1981 deadline for ending ocean dumping in order to obtain a six-month permit to continue ocean dumping.* The plan calls for delivery of most of the city's sludge to a single plant on Wards Island where it would be dewatered and composted. It must then be stored on land or applied to various land sites — mainly park lands.

Program planners acknowledge that this system is unlikely to be completely operative by the federal deadline. In any case, the problem of final disposal remains unsolved because the park land acreage designated for compost is not sufficient for more than a few years' supply and the effects of heavy metals in the compost have not been thoroughly considered.

In the meantime, New York City is leading political lobbying to extend the federal deadline on dumping and pressing EPA for rules changes. *The New York Times* editorially supported postponement of the deadline and city officials believe that a crisis-fired public hue and cry will achieve that purpose on the threshhold of the deadline. The mayor has called for postponement to 1990 — a more realistic target than the original one. But there is little evidence that another decade would end with anything but a similar scenario.

In effect, the federal government has put local governments in a very uncomfortable position and if political history repeats itself, the ban will not work. EPA has left stringent regulations for pretreatment at levels needed in large coastal cities up to the states and localities,

---

*The regional administrator of EPA has informed the city that it has failed to meet some of the scheduled dates for phasing out dumping and that the violations have been referred to the U.S. Department of Justice for action under the Marine Protection Act.

thus shifting the political and economic burdens of raising the costs of doing business to the locality. At the same time, studies of the alternatives to sludge dumping show all of them to have higher financial costs and do not convincingly demonstrate to local constituencies the social and environmental costs of dumping. It is unlikely that the federal government will be able to force localities to increase their disposal budgets appreciably by fiat.

The reduction of dumping from municipal disposal systems would be more feasible if a pricing-incentive system were instituted. Substantial dumping license fees combined with subsidies for sludge tonnage reduction from a baseline, for example, would alter the cost-benefit picture considered by localities. At the same time, concentrated investment in demonstration projects involving pyrolysis, composting, monitoring, and testing wastes would have greater effect than local planning funds spread nationwide.

Regulation of another aspect of ocean dumping is also in jeopardy. Regulation of harbor and river dredging to reduce ocean dumping of dredge spoils has been pursued, also without clear exposition of costs and benefits. The public does not understand the significance of dredge material for ocean pollution. One regional newsletter proclaims, "With so much misunderstood water to sail on, to dredge or not to dredge would appear to be a lesser regional issue than potholes."[10] The technical significance of bioaccumulation testing now being applied by the regulating agencies is under dispute. The regulatory process leading to dredge permit approval or disapproval is unnecessarily cumbersome, involving five federal agencies and up to two years for single-permit review. The economic costs of limiting dredging in port cities have been underestimated. The local interest groups promoting economic development are mounting a campaign for deregulation. One bill has been introduced in Congress to bar the application of bioaccumulation tests. The deputy mayor of New York has stated that the port as an economic mainstay is being "jeopardized unnecessarily . . . by the overzealous interpretation by administrative agencies of their responsibilities under various environmental statutes."[11] The Port Authority of New York and New Jersey painted the picture of abandoned piers and empty warehouses if New York were forced to return to the nineteen-foot depths that nature apparently provides for it (container ships average between thirty- and thirty-five-feet draft).

THE FACTORS OF POLICY STALEMATE

This discouraging picture holds specific clues that are keys to improving performance. This chapter is not directed at analysis of technical methodologies; I leave that to my colleagues in the appropriate sciences. It does attempt to formulate strategies that will strengthen public support and maximize the feasibility of policy choices being implemented.

First, these strategies need to be adapted to the characteristics of water quality issues. These issues are a combination of complex, uncertain, and technical relationships with highly charged political factors associated with multiple use conflicts.

Research and Demonstration

The technical complexities and uncertainties mean that not all the policy questions can be definitively resolved in the short run. Scientific research should be focused on ranked priorities. The cases demonstrate that one priority should be better understanding of cause-effect relationships. To what degree will water quality be improved by specific reduction of each major parameter of pollution? Without technical guidelines of this type, the benefits associated with specific costs cannot be calculated. Conclusions on effects of pollution control must be drawn before definitive proof is in. This requires definitions of unacceptable levels (end points) for the different parameters of pollution and linkages to specific sources.

Measurement and monitoring activities also need nationwide expansion and coordination. Fragmented sampling and research by four levels of government agencies, industry and academic groups are not producing uniform conclusions with comprehensive coverage. New York, New Jersey and Connecticut, and their joint Interstate Sanitation Commission do not have uniform classification of waters or uniform standards for the parameters of pollution. All of the Section 208 planning endeavors suffered from inadequate basic data. Without the data and without water quality priorities, some plans simply listed everything that might be done, far exceeding the willingness of most public officials on local levels to endorse them. There are instances where prioritizing on the basis of better data and cause-effect linkages could identify one or two major actions (e.g., controlling ammonia discharges from a single plant) that would improve water quality more than expenditure of hundreds of millions on waste treatment plants. Similarly, prioritizing on the basis of cause-effect relationships might in some instances reduce emphasis on secondary treatment until sludge disposal alternatives are tested and selected. Where cause-effect relationships remain uncertain, at least risk analysis with probabilities weighted by potential effects could aid decision-making.

Scientific progress in these areas needs to be translated into popular terminology. For example, the policy implications of the Crystal Mountain workshop on "Assimilative Capacity of Coastal Waters"[12] should be drawn in clearly worded issue papers that can be injected into public debate. Communication and public education should be viewed as integral purposes of research and planning.

The state of knowledge on newer techniques of pollution control also implies that demonstration projects should be emphasized. One case in point would be a consolidated regional pyrolysis project designed to test both economic and technical feasibility, and to work out the problems being encountered in local projects such as the Suffolk County plant. Water quality improvement needs more experimentation and evaluation, with national support and dissemination of results.

Justification of Goals

Perhaps one of the most difficult challenges crucial to maintaining momentum for water quality improvement is developing public consensus on realistic goals. Water quality management requires resolution of multiple use conflicts. This is fundamentally a political process, albeit one that should draw on high input of data and technical analysis. Up to this point, national clean waters policy has been absolutist, aiming for zero discharge. The goals that were expressed in 1972 are not realistic. More important, because they did not consider the multiple use conflicts, they did not forewarn us about the political and economic obstacles. We must cease underestimating what to many interests are the *benefits of pollution*. They include cheap industrial processes, port efficiency for shipping, reduced tax burdens for municipal disposal. I use the phrase "benefits of pollution" mainly to induce political realism in my audience. These factors should, of course, be accounted for as costs of pollution control.

One of the major public policy difficulties is that the benefits of pollution control are diffuse and long-term. Public health benefits are widely spread and not discretely visible. Future opportunities for recreation and waterfront revitalization will have long-term effect only when coupled with appropriate land management and development efforts. Aesthetic improvements induce more immediate appreciation; the public tends to be more concerned with floatables than with heavy metals. Aesthetic perceptions often have little connection, however, to the actual parameters of pollution. Waters stirred up by storms look too dirty to swim in to a public affected by rumors of pollution; but potentially toxic pollutants are underemphasized by local officials who wish not to discourage the tourist trade. Fundamental protection of ecological balance competes weakly with short-term economic gain except when specific crises occur or are threatened.

So the benefits of pollution control are diffuse and long-term, but the costs of pollution control are focused and short-term. Their burden is distributed to individual companies, municipal budgets, and site neighborhoods. The interests on whom the costs fall are better organized for political action; water quality goals that lack both realism and cause-effect justifications are vulnerable to their attacks.

Environmental groups are stronger on the national level than within localities. Economic interest groups are often stronger on the local level. Municipal and state officials have to be concerned that industry, taxpayers, and port activities will move away from the local costs of pollution control. Therefore, insofar as the federal government delegates regulatory authority and project costs to state and local levels, it weakens the chances of success. In addition, environmental groups have been more active on issues of air quality and inland waters than on issues of marine pollution; even they need public education about the values of ocean preservation.

Governmental Fragmentation

The cases illustrate that concerns for the ocean are subject to a high degree of governmental fragmentation. The more agencies sharing responsibility for water quality improvement, the less the concerted power exercised and the greater the opportunities for opposing interests to stalemate policies.

Water quality concerns are scattered among five federal departments (Department of Commerce — National Oceanic and Atmospheric Administration; Environmental Protection Agency; Department of the Interior — Fish and Wildlife Service and Bureau of Land Management; the Corps of Army Engineers; the Coast Guard); several agencies in each state; municipal and county agencies; and Section 208 planning agencies. Ambiguous assignment of responsibilities, bureaucratic in-fighting and buck-passing have hindered implementation of water quality improvements.

The literature has repeatedly pointed out ways in which intergovernmental finance patterns skew decision-making. Reliance on federal capital grants and tax exempt bonds, for example, reduces the short-term local construction costs of large-scale treatment facilities by shifting the burdens to the national taxpayers. New technologies, alternative sludge disposal systems, control of products flowing into disposal systems, and reduction of combined sewers suffer by comparison because larger portions of their initial costs fall to the implementing decision-makers. Present financing patterns also put at a disadvantage such tasks as the training of water quality planners, public education and communications, demonstration projects of national significance, water quality monitoring and regulatory enforcement.

Financial Incentives

Policy analysis for improvement of water quality should consider closely the potential of financial incentives which are a means of inducing changes in multiple use patterns. They are not compatible with the absolutist approach to zero discharge. But they are needed to reach intermediate goals if we are to minimize cumbersome regulatory processes. They could be used to induce municipalities to pursue alternative sludge disposal methods. Fees for dredge spoils dumping could also be designed to ration dredging according to economic importance (e.g., control of port silting). Another application of incentives might be to induce industrial pretreatment of wastes that flow into municipal disposal systems. Fee structures might be coupled with tax law mechanisms to ease the burden on small marginal industries of local importance (e.g., electroplating, printing and food businesses in the New York region). Accelerated tax write-offs and use of tax exempt, small-scale industrial development bonds are well tested methods which shift some of the burdens to the national taxpayer. Other industries — chemicals and petrochemicals, for example — can bear the burdens.

However difficult to design, an equitable and effective system of incentives could have large payoffs. Industrial contributions of heavy metals to coastal city disposal systems range from one-third to three-quarters. Carrots must be combined with sticks, however, and public financial or tax subsidies to private industry should be limited to cases where severe dislocation to local economies might result from enforcement of high pretreatment standards.

> The underlying principle in providing modest amounts of assistance is to help overcome possible short run economic hardship which a firm might suffer because of pollution controls imposed by government agencies. . . . There is no justification for wholesale or long run departure from the basic principle that firms [and their products] should bear the social costs of their pollution.[13]

A FRAMEWORK FOR PLANNING

A framework of policy planning for water quality improvement that can mitigate the problems cited here must be comprehensive but simplified in order to bridge the gap between technical analysis and calculations of public interest. Comprehensive means that it should encompass a wide range of technical, social and environmental factors. Simplified means that it must accommodate the constraints on quantification given the present state of knowledge and that it cannot provide holistic, definitive answers across the board. In order to aid public choice under conditions of uncertainty, it should facilitate successive comparisons among policy options and clarify the objective factors that underlie multiple use conflicts.

Cost-benefit analysis as it has been applied to river basin and flood control projects has tended to be too narrow, encompassing mainly financial costs and specific water use benefits (water supply, flood control, recreation, hydropower, fishing and navigation). Social impacts and indirect environmental impacts tended to be only marginally considered.[14] Environmental impact statements, on the other hand, have attempted to catalogue a vast range of quantitative and qualitative data, but they are also usually focused upon single project designs. The literature on environmental impact and cost-benefit analysis is vast. I am oversimplifying the assessments here in the interests of brevity. The crucial point is that neither has been successfully used to compare a wide range of policy options (including the option of nonaction). Nonaction may be unacceptable to most of us here. But its costs and benefits need to be assessed in order to convince segments of the public that remain unconcerned. Finally, two crucial dimensions need to be added to the framework for water quality improvement planning — a time dimension and a distributive dimension.

The Time Dimension

One of the central purposes of planning is to extend considerations further into the future than do the ordinary political decision-making processes. The record of forward-looking planning in the United States is a dismal one, beginning with the political defeat of natural resources planning in the 1930s. The normal perspectives of political officials are geared to short-term electoral cycles, overemphasizing quick results. The political system discounts long-term costs and benefits. With clarification through planning and induced public support, however, long-term impacts can be translated into short-term political costs and benefits by the actions of interest groups or by shifts in public opinion.

The time frame provided for Section 208 plans was inadequate.[15] Even if sufficient data had been available and staff resources had been adequate, the two-year plan completion deadline did not permit forward-looking analysis but put emphasis on techniques that could be implemented in the short run and their near-term costs.

Ecological degradation and its prevention can only be objectively analyzed by considering "a flow of utility over time."[16] In many instances, a view of the future balance sheet will improve the discernible net benefits of pollution control policies, both by emphasizing future environmental costs of inaction (e.g., projected sludge volumes of heavy metals in the year 2000) and by identifying future economic benefits (e.g., waterfront redevelopment) that will offset short-term economic costs. Current capital alterations must be viewed in terms of their impact on future income flows, a concept familiar in business planning.

Viewing costs and benefits in a time stream can also help to include opportunity costs into the calculations, particularly the values of foregone opportunities for high quality marine and waterfront environments.

> High expenses will be involved, certainly at the beginning, and therefore a careful analysis of the reallocation of communities' resources and trade-offs in costs . . . will be called for. There is good hope, however, that these costs will decrease as national, large-scale systems become operative. Furthermore, the [future] costs of not providing environmental control projects could be staggering.[17]

Finally, the time stream helps the planner cope with the difficult problems of predicting and calculating needs for reducing the chances of major accidents or marine environmental crises.[18] Recent work on risks of nuclear power plant accidents provides some insights. The particular causes of specific accidents in highly complex systems have low probabilities of precise recurrence. But in the long-term, high probabilities that accidents of varying precise causes will occur are associated with the proliferation of "highly interactive, complex, high risk systems." This is the formulation of Charles Perrow.[19] It is particularly applicable to oil spills, overflow of sewage solids and floatables, and accidental release of toxic substances. The implications of this concept mean that it is less important to know the precise factors in a given accident and to predict their specific recurrence, than to assess long-term risks on the basis of the volume of accident-prone technologies and practices.

The Distributive Dimension

The planning framework should also encompass a distributive dimension. The various costs and benefits of pollution control options fall to different groups. If multiple-use conflict resolution is to be rational, it must take cognizance of the expected impacts on various segments of the population. All the policy options shift burdens to some extent and clarification of these shifts is essential to effective policy planning. For example, Ruth Mack has formulated an approach to weighting the utility generated by a dollar of advantage for a poor man as greater than that for a rich man by conceptualizing impacts in terms of the percentage of income rather than absolute dollars.

In additon, the planning framework must provide understanding of the rough trade-offs between national and local economies, private and public sectors, coastal and upstream residents, various water using and real estate-based businesses, and so forth. It is not possible to quantify precisely the trade-offs between losses to specific businesses and aesthetic benefits to coastal residents and visitors. But we can at least range them in a planning framework that identifies the trade-offs and provides an equitable base-line for bargaining.

CONCLUSION

How can we persuade politicians to give greater weight to long-term, diffuse issues such as marine pollution? One way is to provide information that raises public consciousness. Even before scientific results give us specific proof, even while cause and effect relationships remain uncertain, we can find ways to open up public discussions about future consequences on public health, on the food chain, on recreation and aesthetic resources. For example, test results from the deepwater dump site number 106 for chemical wastes are ambivalent, but we know that the acids and heavy metals dumped there remain in the ocean, somewhere. The uncertainties of where they go and in what concentrations should be clearly stated. Uncertainty means future risk and future risk is raised by increases in volumes dumped. Therefore, we should be concentrating on the sources — on incentives and technologies for reducing industrial dependence on the harmful materials in the first instance.

Present regulations can translate long-term risks into short-term costs and thereby change behavior. This process is beginning for land-based disposal of toxic wastes. Government regulations are raising the costs of disposing of acid wastes. Several major industries have responded by changing production processes to reduce the use of acids.

Uncertain scientific results should not be used to justify the status quo. They should be used to estimate risks and to reduce them. However preliminary, existing data certainly justify dramatic reduction of the use of cadmium in industry, for example, and hard line approaches to movement of dredge spoils high in PCBs.

In addition to public education and behavioral incentives, mobilization of groups is a key strategy for translating long-term issues into short-term political pressures. The rise of environmental protection groups has demonstrated this phenomenon with respect to conservation of land and to air quality. They have been less active with respect to marine pollution partly because of the less clear exposition of consequences.

Finally, the public officials and scientists who work with government need to develop planning frameworks that are much more persuasive than the work produced by the "208" water quality planning.

The tables that accompany this chapter represent only a rudimentary outline of work that has yet to be done. For them to be filled in and refined, scientific data on the most important parameters of pollution must be further developed and economic calculations on impacts must be made. The approach outlined in the charts attempts to incorporate both the time and the distributive dimensions of marine pollution impacts and alternative policies for reducing them. It attempts to translate effects of pollution into impacts on social groups: on swimming opportunities for inner city children, on shorefront economies, on nutrition and health, etc. Pollution must be so translated from chemical components into social impacts in order to relate it to the public interest.

TABLE 1.1

INTERIM TEN-YEAR PLAN (one of several options)

(This framework should be filled in with estimated water quality effects relative to defined tolerances and 1975 baselines. Expected improvement of environmental quality is indicated by (+)).

	BOD	BACTERIA	HEAVY METALS	TOXIC ORGANICS	FLOATABLES	SUSPENDED SOLIDS	ACID WASTES	TREATMENT RESIDUALS
Municipal sludge dumping continued through 1990			–			–		
Average annual dredge spoils reduced by half from 1975 baseline			+	+	+	+		
Increase in secondary treatment capacity relative to total disposal volume to 80%	+	+			+	+		–
Specified levels of industrial pretreatment regulation to achieve identifiable load changes in municipal systems			+	+			+	
Elimination of specific direct industrial discharges upstream			+	+			–	
Seasonal skimming and disinfection of storm outlets and combined sewer overflows		+		+	+	+		
Capture of 90% combined sewer outlet by 1990		+		+	+	+		
Product control (low lead gasoline, no-phosphorous detergent, etc.)			+	+				

TABLE 1.2

IMPACTS OF TEN-YEAR PLAN: YEARS ONE TO THREE

(Quantification should be possible in most categories but value ranking will be by judgmental index)

GROUPS OF INDIVIDUALS AFFECTED	EFFECT	DESIRABLE (+) or UNDESIRABLE (−) IMPACTS PRODUCED			
		Financial and Economic	Social	Environmental	Crisis Risks
Neighborhoods of new plant sites	Dislocations – can be mitigated N people	Small business losses and gains balanced	(−) Neighborhood disruption	(−) Construction refuse	
Labor at construction sites	N jobs	(+) $			
Large industries: chemical & petrochemical, other (specify)	Pretreatment regulations	(−) $ (Capital Costs)			
Small manufacturing & processing	Pretreatment regulations	(−) $ (Capital Costs)			
Port & shipping businesses	Dredge reduction	(−) $ (Fees and Loss of Volume)			
Recreation & waterfront businesses, including real estate	Minor effects in this time period				
Finfishing industry	Opening some river fishing	(+) $			

TABLE 1.3

IMPACTS OF TEN-YEAR PLAN: YEARS FOUR TO TEN

GROUPS OF INDIVIDUALS AFFECTED	EFFECT	DESIRABLE (+) or UNDESIRABLE (−) IMPACTS PRODUCED			
		Financial and Economic	Social	Environmental	Crisis Risks
Neighborhoods of new facilities					
Labor in pollution control operations	N Jobs, increment	(+) $			
Large industries chemical & petrochemical other (specify)	Pretreatment regulations	(−) Decreasing impact on annual income			
Small manufacturing & processing		(−) $ Decreasing impact			
Port & shipping businesses	Dredge reduction	(−) $ Fees			
Recreation, waterfront, real estate & commerce	Land value increases	(+) $	(+) Neighborhood upgrading		
	commercial expansion: initial investments & beginning returns	(+) $			
Finfishing industry	Increase in fish propagation	(+) $			(+) Reduced risk of fish kills
Shellfishing industry	Reopening beds with estimated N yield	(+) $			
All ocean users	Increase in sludge volume and decrease in dredge spoils		(+) Aesthetic enjoyment, scientific interest	(−) Increased disturbance of ocean bottom at dumpsite	(−) Increased risk of beach fouling

TABLE 1.3 *Continued*

	Increase in dissolved oxygen Use density up Decrease in toxics			(+) Aquatic life balance improved (−) Including recreation wastes and treatment residuals (+) detoxification	(+) Reduced risk of food chain contamination
Inner city families	Beach capacity doubled Fish food supply up	(+) Beach related small businesses	(+) Health (+) Man/days recreation (+) Nutrition, well-being		
Regional residents	Outdoor recreation capacity expansion Improved attractiveness of regional location Fish food supply and prices	(+) $	(+) Man/days recreation (+) Nutrition, well-being		
National taxpayers	Tax subsidies Enforcement cost increases	(−) $ (−) $			
Municipal taxpayers	Operating and maintenance Capital costs, P. C. Waterfront development Business and property tax effect	(−) $ (Increasing) (−) $ (−) $ (+) $ (20–30 Year Span)			
Domestic feepayers					

TABLE 1.3 Continued
IMPACTS OF TEN-YEAR PLAN: YEARS FOUR TO TEN

GROUPS OF INDIVIDUALS AFFECTED	EFFECT	DESIRABLE (+) or UNDESIRABLE (−) IMPACTS PRODUCED			
		Financial and Economic	Social	Environmental	Crisis Risks
Shellfishing industry	Minor effects in this time period				
All ocean users	Decrease in visible aspects of pollution		(+) Aesthetic Enjoyment	(−) Waste from Private Boats and Other Recreational Uses	(+) Reduced Risk of Bacterial Infection
	Decrease in bacteria		(+) Increase Confidence in Safety	(+) Aquatic Life	
	Decrease in BOD (still exceeding standards)				
Inner city families	Opening of local beaches reachable by public transit		(+) Man/Days Recreation; Youth Activities		
Regional residents National taxpayers	Capital Costs	(−) $			
	Tax Incentive costs	(−) $			
Municipal taxpayers	Capital costs	(−) $			
	O & M costs	(−) $			
Domestic feepayers	Water related business taxes	(−) $			

Notes

[1]PL 92–500 (86 Stat. 816).

[2]*The New York Times.* January 19, 1979, 8:1.

[3]Marx, W. 1967. *The frail ocean.* New York: Coward McCann, p. 7.

[4]National Oceanic and Atmospheric Administration. 1980. *Proceedings of a workshop on assimilative capacity of U.S. coastal waters for pollutants.* (Crystal Mountain, Washington, July 29–August 4, 1979). Edward D. Goldberg, editor. Boulder, Colorado.

[5]Comptroller General of the United States. December, 1979. *Water quality management planning is not comprehensive and may not be for many years.* Washington, D.C.: CED 78–167.

[6]Citizens Advisory Committee, New York 208 Water Quality Management Program. April 4, 1979. *CAC comments on 208 water quality management plan.*

[7]Marine Protection, Research and Sanctuaries Act of 1972 as amended (33 U.S.C. 1401).

[8]Comptroller General of the United States. January 21, 1977. *Problems and progress in regulating ocean dumping of sewage sludge and industrial wastes.* Washington, DC.C.: CED 77–18.

[9]*Ibid.,* p. 20.

[10]Port Authority of New York and New Jersey. December 1979. *Report,* Vol. 1, No. 3, p.1.

[11]Statement by Robert F. Wagner, Jr. April 1980. *Transportation Report.* New York Chamber of Commerce and Industry. p. 7.

[12]*Proceedings of a workshop,* Crystal Mountain, *op. cit.*

[13]Institute of Public Administration. February 1975. *Industrial incentives for water pollution abatement.* Report to U.S. Department of Health, Education and Welfare, p. 89.

[14]Institute of Public Administration. 1972. *Northeastern United States water supply study.* Report to the Department of the Army Corps of Engineers, 1972.

[15]Comptroller General of the United States, 1979, *op. cit.*

[16]Mack, R. P. 1974. Criteria for evaluation of social impacts of flood management alternatives. In C. P. Wol, ed. *Social impact assessment.* Environmental Design Research Association. I have borrowed liberally, and with permission, from ideas in Mack's "E-Model" in outlining the planning framework.

[17]Grava, S. 1969. Urban planning aspects of water pollution control. New York: Columbia University Press, p. 165.

[18]Saila, S. B. 1978. Some postulated associations between biological conditions in the New York Bight and marine environmental crisis. Unpublished. University of Rhode Island.

[19]Perrow, C. *TMI: A normal accident.* unpublished manuscript.

Discussion

Kester: Would you suggest that an effort be made to enhance public education in the area of marine pollution or direct it more toward environmental management?

Walsh: I think it should be directed toward environmental management but the public education dimension should not be ignored. Some of the more general groups that have been very active in air pollution could be effective in marine pollution control if they were brought into the picture.

Comment: You spoke of pollution in terms of the costs being concrete, specific, short-term, and the benefits being long-term and diffuse. I understand the finance side, but I am curious about the political side. How can one make politicians more responsive to longer-term considerations when they are up for election every two to four years and since they are subject to interest group activity? It is fairly obvious that those people whose interests will be specifically impinged upon will be those who are most active; those who have a diffused interest in activity will sit back. I agree with what you are saying about finance, but how do you turn it into political reality, given the system as it exists?

A. Walsh: What typically happens in our society is that you have to take the discounted long-term diffuse benefits and translate them into short-term political cost benefits by organizing the interest groups who make that concern active and immediate to the political officials. That is why I think you have to go beyond the environmental management groups to the general good government groups, the media, and the various and sundry generalized groups. If we can get them excited about marine pollution controls (which I think we have failed to do) we can begin at least to build up some of the short-term pressures in the political system. This has happened in air quality in many areas.

Goldberg: I appreciate your mood; the fragmentation of our agencies or groups in the United States has been detrimental to the bulk of rational environmental policy.

Another problem relating to that, which you did not mention, and which concerns me, is that in contrast to some other countries, we really have no method of getting a scientific consensus in the United States. For example, in the United Kingdom the expansion of nuclear deposits in the plant at Windscale is based upon the Park Commission Report, which I think is one of the most important environmental documents produced in the past decade. An agreement was developed which had an economic basis as well as a scientific one. In the United States we have a problem now of the ultimate disposal of radioactive waste. We not only have fragmentation among government groups as to who is responsible for the final decision but also we have no scientific consensus.

Comment: I was interested in your comment that there should be a followup to the Crystal Mountain workshop on assimilative capacity. From your point of view, can you tell me what sort of follow-up you mean? Is it possible that the document has gone as far as it can go?

Walsh: That may be the case, but by follow-up I would be hoping that the 208 water quality management planning, for example, would be aimed at developing public understanding and local public policies. They should be taking documents like the reports from the conference and reporting them in a format that would bring the implications to bear on local policies. I am really talking about putting documents in a framework that could tend to communicate to the public what the implications of the present state of research is as it progresses.

CHAPTER 2

The Oceans as Waste Space

EDWARD D. GOLDBERG
Scripps Institution of Oceanography
United States

It has been about thirty years since scientists recognized that renewable resources of the ocean could be placed in jeopardy as a result of man's activities. This awareness came about in the early 1950s when collectives of scientists in the United States, the Soviet Union and England, among other countries, became concerned about the uncontrolled release of artificial radionuclides to the atmosphere and to the ocean. As a consequence of these concerns, their deliberations provided guidelines for the management of radioactive wastes in this developing period of nuclear energy.

We have not only a scientific basis which will identify materials that could result in the loss of marine resources but also have been alerted by a number of catastrophes. The textbook example is the Minamata Bay episode, where the Chisso Chemical Company, located on Minamata Bay, released wastes from the production of plastics into the semi-enclosed marine basin. The discarded materials contained manganese, sulfates, mercury, and organic matter; but in this mixture methyl mercury chloride evolved and entered the marine food chain. The methyl mercury chloride poisoned the people who ate large quantities of fish and shellfish, primarily fishermen, their families, and even their pets. This epidemic was first recognized in 1953. The culprit in the case was not inorganic mercury or elemental mercury, which are involved in classic toxic episodes, but was this very unusual neurotoxin, methyl mercury chloride.

There is also the Kepone episode in the James River, into which a chemical company promiscuously released wastes from its pesticide producing plant. The powerful biocide entered the marine food web and resulted in the closing of some commercial fisheries. The catastrophe was not identified by a marine related event, but it was recognized by the impact of the chemical upon the plant workers who came down with neurological diseases. Then there is the Bedford Bay, Massachusetts incident, where polychlorinated biphenyls have entered the estuarine zone through discharge from an industrial operation. The levels in fish reached unacceptable values for human consumption. Herein, we have another example of a fishery being closed.

The Minamata Bay, the Bedford Bay, and the James River episodes have resulted in losses to society of at least hundreds of millions of dollars, an amount which surely dwarfs all of the research funds that have gone into marine pollution, at least in the U.S. and Japan over the last few years.

We've had other cases where the scientific community has identified a pollutant. The unacceptable effects of DDT upon non-target organisms were recognized by two U.S. scientists, C. Cottam and E. Higgins of the Fish and Wildlife Service, who predicted exactly what would happen early in 1946 just after its introduction as a biocide.

These concerns with marine pollution have led us to identify what we are protecting. The primary efforts have usually involved public health. This developed from our initial assessments of the dangers of artificial radioactivity in the environment where our citizenry receives ionizing radiation through the consumption of seafoods or through exposure to contaminated sediments in coastal areas. In a much more sophisticated vein, we have sought to protect ecosystems. Many northern hemispheric countries have banned production and use of such toxic chemicals as the polychlorinated biphenyls and DDT whose deliberate or inadvertent entry into the environment has jeopardized the well-being of marine and terrestrial organisms. We have also been concerned with protecting the non-living marine resources such as recreational areas, transportation routes, and aesthetic pleasures.

The sites of pollutant releases are primarily in the northern hemisphere where societies have developed that have utilized large amounts of materials, some of which enter the marine environment and have deleterious effects upon its resources.

Our concerns with materials entering the ocean are not only with pollutants, but are a part of a much bigger problem — the ability of the ocean to accept a portion of the benign wastes of our society. Each year the human population produces somewhat over three billion tons of solid and liquid wastes which corresponds to a cube, a kilometer or so on edge. This excludes the products from the annual burning of fossil fuels, which turn out to about twenty billion tons of carbon dioxide. Thus, we have about three billion tons of forest products, minerals, and food products, which pass through our society. The key words here are "pass through" because human society today is a disposing society, not an accumulating society.

We have three optional sites for disposal of these materials. We can put them into the atmosphere, into the ocean, or upon land. The caveat for ocean disposal is that *no renewable resources are lost*.

My springboard in considering this problem will be the history of our studies with marine pollution, beginning about thirty years ago, in the early 1950s. These researches will perhaps guide us to more reasonable policies in using the ocean as a waste space.

What are the key points we've learned in marine pollution investigations? The radioactive pollution problem is a rewarding point of entry as it was recognized in the early 1950s. A very profound expression of the mood of the period was made by Revelle and Schafer (1957).

> Among the variety of questions generated by the introduction of radioactive materials into the sea, there are few to which you can give precise answers. We can, however, provide conservative answers to many of them, which can serve as a basis of action pending the result of detailed experimental studies. The large areas of uncertainty respecting the physical and chemical and the biological processes of the sea, leads to restrictions on what can now be regarded as safe practices. These will probably prove to be too severe, when we have obtained greater knowledge. It is urgent that the research required to formulate precise answers be vigorously pursued.

As a consequence of policies made by individual sovereign nations, the United States, the United Kingdom, and the Soviet Union, among others, we have maintained, I submit, oceans and atmospheres free of harmful levels of radionuclides.

Perhaps the most important control methodologies were developed in the United Kingdom, where today the world's largest discharges of radioactive wastes are made to the Irish Sea from the Windscale nuclear facilities. The British have developed the "critical pathways" approach.

Each year the British determine and publish the amounts and types of radioactivity introduced to the sea. Their objective is to protect human health. They seek answers to the question "How can radionuclides released to the environment be returned to human society and jeopardize their well being?" The British scientists argue that for any given radionuclide, among the many released to the environment, only a few will return to human society and endanger public health through direct exposure and through the consumption of seafood. An example of this approach involves the regulated releases of the radionuclide, ruthenium 106, from the waste pipeline into the Irish Sea at Windscale. Ruthenium enters the marine foodchain in algae. In one plant, *Porphyra,* it is remarkably concentrated; the *Porphyra* is harvested by some societies as food. The people of South Wales convert it to a pudding called laverbread. It is consumed by a small population, some of whom eat two or three hundred grams a day. *Porphyra* harvested near Windscale has contained this radioisotope, Ru-106, which can endanger public health following ingestion in food by its concentration in the gastro-intestinal tract.

The British, in regulating the releases of Ru-106 to the environment, have attempted to protect the highest consumers of laverbread. The maximum permissible exposure of the most avid consumers is based upon data of the International Commission on Radiological Protection (ICRP). This organization has considered what are the upper limits for body burdens of Ru-106 which provide no apparent danger to the health of an individual. The amounts of a given radionuclide entering the Irish Sea from Windscale are related to the maximum amount of radiation which would be received by the person receiving the greatest exposure. For the period 1970–1974, the consumers of laverbread would have consumed 0.2 to 33% of the recommended ICRP limits. In a similar way, the British have protected a single fisherman, who spends a large amount of time fishing in an estuarine beach area where there are gamma emitting radioisotopes which have accumulated in the sediments. The radioactivity to which our single fisherman was exposed would have been 7 to 12% of the recommended ICRP limits.

From studies on the entry of artificial radioactivity to the marine environment, where scientists understand the problem, appropriate controls can be formulated such that we can protect human health. This is an important lesson learned during the early period of marine pollution investigations.

A second lesson came from the problems with mercury, initially identified by the Minamata Bay Disease episode. It was in 1953 that the first victims of the epidemic of methyl mercury poisoning were recognized. It was in 1960 that mercury was discovered to be the element responsible for the disease, and it was in 1963, about ten years after the first victims were recognized, that methyl mercury chloride and methyl mercury were recognized as the culprits in the disease. Methyl mercury poisoning is now known in the medical profession as the Minamata Bay Disease. The studies on mercury by environmental scientists led to the knowledge that the normal form of mercury in fish is methyl mercury chloride.

In the late 1960s and early 1970s, the Swedish government also became aware, through their scientists, that they also had an environmental mercury problem. This was provoked through the dissemination of grain seeds coated with mercury organics, used as anti-fungal agents. Their problem was somewhat similar in nature to the methyl mercury problem that the Japanese had in Minamata Bay. The Swedish scientists recognized that mercury poisoning, through these organic materials, could be widespread, especially through the consumption of fish or grain eating birds. They asked the question "What is an acceptable level of methyl mercury intake to a human, through the consumption of fish and birds?" Fish eating was the more important problem. Utilizing the lowest mercury levels in Minamata Bay victims, the known residence times of methyl mercury in humans, and applying a safe or contingency factor of ten, a daily uptake of methyl mercury with minimum risk was evaluated. For those fish eaters that consumed at least one fish meal a day, the maximum permissable level in fish was evaluated to be 0.5 ppm (wet weight).

This limit protects the heavy fish eater in Sweden and in many countries, since many countries adopted this limit of a half part per million. The heavy fish eater is a person who eats three or four hundred grams per day, a single meal.

The important lesson the mercury tragedy taught is that scientists can reach an understanding of a critical pollution problem in the coastal zones and can propose remedial actions in terms of decades. The mercury problem started in the 1950s and the solutions or the options for various political actions were formulated in the early 1970s.

Another lesson from the studies in marine pollution which may help in assessing the oceans as waste spaces for some of the unwanted materials of society came from the perils of DDT. This pollution episode began in 1946. Two U.S. Fish and Wildlife scientists (Cottam and Higgins) said at that time:

> From the beginning of its wartime use as an insecticide, the potency of DDT has been the cause of both enthusiasm and great concern. Some have come to consider it a cure-all for insect pests; others are alarmed because of the potential harm. . . . DDT, like every other effective insecticide or rodenticide, is really a two-edged sword; the more potent the poison, the more damage it is capable of doing. . . . The most pressing requirement is a study to determine the effect of DDT as applied to agricultural products on the wildlife and game dependent upon an agricultural environment. About 80% of our game birds, as well as a very high percentage of our non-game and insect eating birds and mammals, are largely dependent upon an agricultural environment. In such places the application of DDT will probably be heavy and wide-spread; therefore, it is not improbable that the greatest damage to wildlife will occur there. Because of the sensitivity of fishes and crabs to DDT, avoid as far as possible the direct application to streams, lakes, and coastal bays.

In 1946, these two scientists raised a flag of warning; subsequent events clearly confirmed their observations and predictions. One of the more publicized impacts upon ecosystems involved the reproductive failures of the pelican population on the Anacapa Islands off the California coast from 1969 to 1972. Apparently, the accumulation of DDT in their foods from the sea initiated the problem. There is also the case on adjacent islands of sea lions aborting their young, allegedly because of high body burdens of DDT and its degradation products.

The lesson we learned from DDT in the environment is a very positive and important one. A government can restrict the release of a pollutant to the environment, not based upon protecting human health, but upon protecting ecosystems. Many chlorine containing pesticides, as well as chlorine containing organic chemicals, like the polychlorinated biphenyls, have been prohibited because of their impact on ecosystems.

Over this thirty year period of an awareness of the measurable impacts of man on the oceans, a large number of polluting materials entering the coastal environment have been identified: heavy metals, radioactive nuclides, chlorinated hydrocarbons, petroleum hydrocarbons, litter, microorganisms, dredge spoils, and a group that has attracted a lot of attention in the last five or ten years, the oxidation products resulting from the chlorination or ozonation of waste and cooling waters. The oxidation products result from the interaction of chlorine (and consequentially bromine) with the organic materials contained within the water. The brominated and chlorinated organics include some carcinogenics like chloroform and require an assessment as to their potential dangers.

Now, simultaneously with the development of the concepts about marine pollution and with the identification of pollutants, we have increased our knowledge about marine science. There has been a dramatic growth in our understanding of ocean chemistry, a part of which has, by the way, come from pollutant studies. For example, the importance of atmospheric transport of organic materials from the continents to the oceans was acknowledged following studies on the dispersal of DDT and the polychlorinated biphenyls. The existence of methylated species of metals and metalloids in sea waters and some of its organisms evolved from the Minamata Bay incidence. We've learned a good deal about the enrichment of metals and organics in the marine biosphere. We've had a minor revolution, or perhaps even a major revolution, in our ability to analyze sea water. In the past five years, the analysis of many metals at the parts per trillion level has finally been achieved.

Now, with this background, we can consider the problem of the three or four billion tons waste moving through the world society each year. Where do we put these materials and still keep our marine resources in renewable states? Our options are limited to atmosphere, ocean, or land disposal. For some substances, combustion and release to the air of the products is most reasonable. I would argue that if we're going to rid ourselves of some noxious chemicals we have on the surface of the earth, like PCB's and nerve gases, combustion of the organic chemicals to benign substances like carbon dioxide, water and hydrochloric acid is an acceptable tactic. Clearly, the atmosphere is the most reasonable place for disposal and may become a conventional technology in waste management.

The lands of the earth are most reasonable to take other wastes. For example, the high level radioactive wastes, i.e., the spent fuel elements from nuclear reactors, should be buried in deep silts where water hasn't been seen for millions of years, like the shield areas.

On the other hand, the oceans would be a most reasonable place to put some domestic, industrial, and agricultural wastes. Clearly, not only must one consider a scientific basis for discharging wastes but also one must look at economic, social, and political issues.

How does one ascertain what amounts of materials can be placed where in the oceans and at what rates? This was the subject of the Crystal Mountain Workshop held in August 1979 under the sponsorship of the Environmental Research Laboratories of NOAA. When one asks about assimilative capacity of a body of seawater for a specific type of waste, one can compare this to a titration that chemists are accustomed to carrying out in their laboratories. The pollution titration is one in which pollutant materials are added to a given body of water and an end point is sought, which is determined by the amount of pollutants that the waters can handle without an undesirable effect or an unacceptable effect, i.e., the loss of a marine resource. A significant and substantial direction that scientists concerned with marine pollution might take in the forthcoming years is the development of end points and their field determinations. We seek measurable quantities in the environment which would indicate when we are close to or have exceeded the capacity of waters to handle waste materials without the loss of resources.

There have already been developed a number of end points in the studies of marine pollution for individual substances and for radionuclides. One example involves the radio isotope Cs-137 released from nuclear energy facilities. The amounts of Cs-137 issuing from the Windscale facility in England are regulated upon the levels in fish such that the highest consumers can eat them without endangering their health. The Ru-106 end point for the algae *Porphyra* is another case in point. And of course, the level of methyl mercury in seafood is another end point which protects humans from the ingestion of these potent neurotoxins.

We have formed a large number of end points for individual organic substances, for many other radioactive materials, and for certain metals. How does one go about formulating end points for the disposal of wastes which may contain many pollutants? This is one of the important problems that are going to have to be looked at in the future. Already, some end points have been proposed. Fin rot in fish exposed to a large number of organic pollutants in sediments might be an end point. In Southern California waters it has been observed that about 5% of fish such as the Dover Sole have fin rot and stomach tumors. The Dover Sole is not a terribly important fish commercially in Southern California, but to ecologists 5% may be too high an incidence to tolerate. This is where one may get into a subjective judgment. Egg shell thinning of bird populations, related to the uptake of DDT, PCBs, and other chlorinated hydrocarbons, could be used as an end point. One of the consequences of the uptake of DDT by birds is an interference with calcium metabolism; the birds then lay eggs with much thinner shells than normal, which are friable and easily cracked. A population decline occurred as a result in the pelicans of Anacapa Islands. Perhaps an end point for the release of chlorinated hydrocarbons in waste might be obtained by measuring the thickness of bird shells. When they start thinning, a problem is identified. The question is then, What amount of thinning, if any, is tolerable?

Perhaps, bivalve health offers an opportunity to obtain end points since some bivalves are quite resistant to impact of pollutants. Yet, on the basis of their ability to accumulate pollutants and to respond through biochemical and histopathological disturbances, they appear most attractive.

The solutions to problems of working with collective pollutants rather than single pollutants will be decisive in decisions with respect to waste disposal. The synergistic and antagonistic effect of pollutants will be involved. Many pollutants are synergistic in their effects. On the other hand, there are cases, well documented today, where one pollutant reduces the impact of another pollutant. A textbook example involved the antagonistic effect of selenium on mercury toxicity.

Finally, I would like to point out that remote sensing is going to play a most important role in our use of the ocean as waste space. Where we wish to make large numbers of observations over large areas over frequent time periods, remote sensing may provide the means to do it.

Over the last thirty years we've developed an understanding of how man's wastes have had an impact on the environment both through rational thinking and catastrophes. Hopefully, this experience will provide to those responsible for managing the environment a better base for decision-making in waste disposal — a scientific facet, which must be combined with economic, political, and social concerns.

Discussion

Comment: Your argument largely appears to be an acknowledgement that we could dispose of wastes intelligently. It seems to me that the vision we have in the American scientific community relative to the nuclear wastes is not portrayed in the story you gave on Windscale because the vision seems to revolve around the engineering hazard and the possibility of engineering error. I would think that the scientific community as a whole might come to a rather quick agreement with you relative to disposing of nuclear waste. But they will not agree with you as quickly on the engineering hazard question.

Goldberg: I do not disagree with you. I think you have stated a problem that haunts us all. How one manages the production of nuclear energy and totally minimizes human error, especially carelessness, is a paramount concern.

Comment: In your discussion of radioactivity, you did not say much about alpha emitters, which would seem to me a most important point. Also, relative to the methyl mercury question, I would like to reinforce what you said. Methyl mercury naturally occurs in fish, and it is found in extremely low concentrations in the coastal waters and even where there is some mercury discharge. In California, for example, we found that a very high level of total mercury in the sediments would be 1%. That is about as high as we have ever found, and of that amount .02–.03% would be totally organic mercury. That is an extremely low amount, and most of it is viewed as entirely natural within the ocean.

Concerning your last point relative to the DDT and PCBs, everyone would agree that we should not put them in the ocean if we can avoid it. The question really is, how do you avoid it? The total output of all Southern California, for both of those two is about one kilogram each per year, and that is in a roughly four billion liters a day. That is a tiny amount. The question is, how do you now discharge it even if you wanted to? And, by the way, that is almost exactly equal to the amount of fallout of those substances into the sea (which is also avoidable, it would seem, but somehow it is not). In some years, after these things have been prevented from ordinary discharge and use, they still show up in the environment. The numbers have become pretty small, and I would like to know why you did not discuss this.

Goldberg: With the regard to the alpha emitters, I did not mention them, but the British do control the amounts of alpha activity entering the Irish Sea. It is primarily plutonium from their reprocessing plant. There is less than 1% of acceptable levels, based on ICRP recommendations, entering the Irish Sea.

With respect to mercury, I am in complete agreement, but this is an important observation. It taught us that the primary form of mercury in fish is methyl mercury.

Finally, about DDT and other forms of hydrocarbons, how do we keep these from going into the ocean? Well, DDT really was not an effective agent in the management of our agricultural efforts. It is well known now that many pests develop resistance to DDT and other biocides. They are being drastically reduced in usage. I think our real concern is with industrial chemicals like PCBs entering the marine environment. Their toxicity has been known since the early 1950s.

Comment: It is also worth noting, that in the case of DDT, as soon as it was recognized what the problem was in about 1971, all the sources were shut down for about a week. It is not as though nobody took action on that. Since DDT was made to kill insects, it is no surprise that it killed mostly microcrustaceans in the sea and upset the balance of things. But everyone tries; it is not as though it is being ignored.

Goldberg: We can protect ecosystems by taking DDT and other such pesticides off the registry when their undesirable effects are identified.

Comment: I would like to raise a couple of questions on your concept of assimilative capacity. First, when you look at the backlog of information that we have and where we have been, we are able to determine the capacities for radionuclides and for, at best, a handful of chemicals such as DDT, PCBs, and nerve gases. But I am reminded of the large volume that the EPA has of supposed toxic materials in a computer listing about a foot and a half high. Our knowledge of the routes, rates, reservoirs, and rescues of those in natural systems are so poorly known that our capability of looking at each of these problem areas is poor.

The next issue concerning toxicity is the determination of end points. The EPA has promulgated marine water fouling criteria, for, I think, sixteen chemicals or compounds out of the some fifty thousand.

Finally, I am concerned about the analogy of the chemists with the titrator trying to determine what is a safe end point. I do not think that is a scientific judgment, to a large extent, but a societal judgment. I am reminded that with petroleum hydrocarbon we can find biological effects; meaningful biological effects at ranges from 5 parts per billion up to 500 parts per million, which is 5 more orders of magnitude. Well, where along that spectrum of snails crawling faster to the killing of every fundulus in your assay, do you set that acceptable limit? I am afraid that you may be offering us false hopes that scientists are going to be able to titrate out and go from blue to red and we are going to be able to determine this for all 50,000 chemicals in our toxic waste chart.

Goldberg: He should have given the talk, not me. I do not think the situation is that bad. There may be 50,000 compounds that we might concern ourselves with. I think, in general, those with highest priority have been identified. I suspect that a couple more "kepone-type" incidents may identify others, but I think we have identified most of the bad actors. We may not have identified the effects of the bad actors. For example, how serious is the production of chlorinated and brominated organics around waste water or cooling water outfalls where the chlorinating and bromination takes place? I just do not know, but at least we can see the problem.

CHAPTER 3

An International Perspective on Global Marine Pollution

MICHAEL WALDICHUK
Department of Fisheries and Oceans
Canada

INTRODUCTION

Pollution knows no national boundaries in marine waters. Limits of national jurisdiction imposed in a geographic sense along a coastline and into waters out to 12 nautical miles for the territorial sea, or to 200n miles for extended national jurisdiction over fisheries or other resources, are meaningless in terms of controlling the movement of pollutants. Effluents discharged by one coastal state can inadvertently defile the waters of another simply through the transport by littoral currents. Ships can directly pollute the waters of a coastal state by deliberate or accidental discharge of oil or other substances into these waters, or indirectly, if such discharges on the high seas are carried inshore by currents. Clearly the oceans can only be protected from pollution by international agreements among countries bordering this vast body of water that occupies 71% of the earth's surface.

There has been a growing demand for protection of the coastal waters of one state from pollution caused by a neighboring state. In some cases, where two neighboring states are involved, it has been possible to negotiate bilateral agreements; in others, where a number of states are involved, it has been necessary to develop regional treaties or conventions. An understanding exists generally among nations of the world that the truly international part of the oceans beyond the limits of national jurisdiction must also be protected from pollution as the "common heritage of mankind." This is where the Law of the Sea comes in.

The international approach to the marine pollution problem generally reflects national attitudes. During the late 1960s and the first half of the 1970s, there was growing environmental concern in western nations, fueled by reaction to a number of incidents of acute pollution, such as Minamata disease in Japan, the *Torrey Canyon* disaster in the English Channel, and more recently, the erosion of the ozone layer, and acidification of lakes. This was the era when the environmental activist groups were most vocal. The public outcry against the type of pollution that was receiving much attention from the media led to political action with strong environmental legislation in countries such as the United States, Canada, Japan, United Kingdom, and in other countries of Western Europe. These national concerns were reflected in various international fora, which were climaxed with the United Nations Conference on the Human Environment held in Stockholm in June 1972. In areas where international collaboration was required to control pollution, conventions were developed at that time. Some of these have been of a regional type, while others have been of a global character, e.g. conventions on prevention of marine pollution by ocean dumping.

The understanding of the effects of marine pollution is far from complete. It is sometimes difficult enough to identify the problems in a given area caused by a specific pollutant, let alone to quantitatively assess the effects. Without this kind of understanding of cause and effect, it is very difficult to predict the environmental and ecological consequences of a given type of pollution. Recognizing the problem of a lack of full understanding and inability to predict effects of pollution, a number of UN agencies, as well as the Scientific Committee on Oceanic Research (SCOR), have developed working groups to examine specific types of pollution problems, to draw some general conclusions based on available information, and to identify the gaps in knowledge.

The needs for marine pollution research can only be satisfied, by and large, through national laboratories, because UN agencies, with a few exceptions, do not have their own facilities for such work. Unfortunately, national laboratories do not always meet the needs identified by UN agencies, partly because there is seldom any obligation to do so, and partly because national priorities may be perceived as being different from the international ones. Under certain circumstances, international laboratories, such as the International Laboratory of Marine Radioactivity in Monaco, may be established to meet a specific need.

Attempts have been made to organize marine pollution monitoring programs under the auspices of UN agencies. These have been difficult to initiate partly because of the high cost, partly because of unavailability of adequate analytical techniques, and often because of the lack of a clearly defined objective. The one successful pilot monitoring project worthy of mention is the Marine Pollution (Petroleum) Monitoring Pilot Project in the North Atlantic, coordinated by the Integrated Global Ocean Station System (IGOSS) under the Intergovernmental Oceanographic Commission (IOC), during the period from 1975 to 1980. It was implemented as much to test the capability of a system (IGOSS) as to actually monitor pollution.

Short term projects with clearly defined objectives have met with greater success under international coordination. There have been a number of successful cooperative experiments to study processes related to marine pollution, such as the project RHENO to study dispersion in the North Sea, under the auspices of the International Council for Exploration of the Sea. In the Baltic, bordering states have successfully collaborated in a number of projects to study marine pollution under regional arrangements. A number of marine pollution research projects have been conducted under the International Decade of Oceanic Exploration (IDOE), an element of the Long-term and Expanded Programme of Ocean Exploration and Research (LEPOR) of the IOC, which have been funded mainly by the U.S. National Science Foundation. There is a recognized need for much more international cooperation to study the pollutant transfer processes in the sea, ecological effects and the fate of pollutants, so that existing conventions and treaties can be better administered and new international agreements can be developed on a sound scientific basis.

This review covers the growth, climax, stabilization, and decline in the international environmental movement during the two decades from 1960 to 1980.

HISTORICAL DEVELOPMENT OF THE INTERNATIONAL OUTLOOK ON MARINE POLLUTION

Recognition of the international implications of environmental pollution had its origin in atmospheric problems. The landmark case was that of the Trail Smelter in British Columbia, for which the arbitral tribunal in 1941 ordered Canada to pay compensation to the United States for damages caused by fumes of a privately-owned Canadian smelting company. The crops in eastern Washington State were shown to be damaged by sulfur dioxide emitted by the smelter, which was transported to Washington State by winds. This decision (". . . Under the principles of international law . . . no state has the right to use or permit the use of its territory in such a manner as to cause injury by fumes in or to the territory of another or the properties or persons therein. . . .") established the legal precedent that one country should not have the right to pollute the environment of another, even though the polluting state may innocently release the offending substance into its own environment and natural forces transport it into the environment of another state.

The widespread international concern about the ill-effects of radioactive fallout from nuclear weapons tests in the 1940s and 1950s led to the 1963 Treaty Banning Nuclear Weapons Tests in the Atmosphere, in Outer Space and Under Water. The First Law of the Sea Conference, held in Geneva in 1958, developed a number of conventions with important provisions on control of marine pollution. The Convention on the High Seas, 1958, for example, considered the problem of oil pollution, wastes from exploration, and exploitation of the sea-bed and its subsoil, as well as radioactive wastes.

The problem of oil pollution in the marine environment, especially that arising from tanker traffic, has been recognized for many years. As long ago as 1926, the United States hosted a conference to formulate an international convention on the prevention of pollution by oil. It failed to develop an acceptable agreement. Oil pollution from ships in international waters remained uncontrolled until the Convention for the Prevention of Pollution of the Sea by Oil, 1954, was brought into force in 1958. It is through this convention and its amendments of 1962 and 1969 that pollution of the sea by oil released from ships is controlled. The Convention on the Prevention of Pollution from Ships (MARPOL), 1973, was developed to replace the 1954 convention but it is not yet in force. The Inter-Governmental Maritime Consultative Organization (IMCO), formed in 1959, administers all international matters related to ships, including marine pollution arising from their operation and international conventions developed to control it.

Out of the 1972 UN Conference on the Human Environment emerged the United Nations Environment Programme (UNEP) with headquarters in Nairobi, Kenya. An action plan was developed for UNEP, and during the first few years of its existence, UNEP supported environmental projects of other UN agencies. Its role in the marine environment has accelerated in recent years with the development of its Regional Seas Programme, which now includes 10 marine regions (fig. 1). In an effort to assess the state of the global environment in the decade after the 1972 Stockholm Conference, UNEP has undertaken the preparation of a comprehensive technical document focussing on changes (positive and negative) since 1972. The marine environment will receive substantial coverage.

The Intergovernmental Oceanographic Commission of Unesco was one of the UN agencies to become involved comparatively early in marine pollution. It formed the Working Group on Marine Pollution, which held its first and only meeting in Unesco, Paris, August 1967. When the interagency Joint Group of Experts on the Scientific Aspects of Marine Pollution (GESAMP) was formed in 1969, all the UN agencies having groups dealing with marine pollution agreed to dissolve them

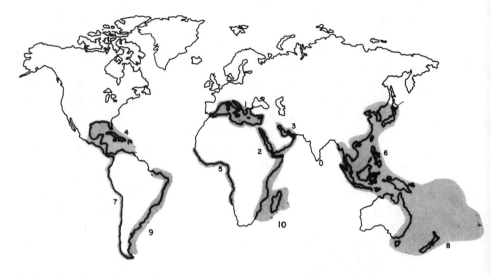

The land and seaward limits to the regional seas indicated in this map are merely illustrative. Definition of the boundaries is the responsibility of the Governments concerned. Action plans are in effect for (1) the Mediterranean, (2) the Red Sea and Gulf of Aden, (3) The Kuwait Action Plan Region. Action plans are in various stages of preparation for (4) the Caribbean, (5) West Africa, (6) East Asian Seas, (7) The South-East Pacific, and (8) the South-West Pacific.

Fig. 3.1   Global coverage of the UNEP Regional Seas Program, (Courtesy of the Regional Seas Program Activity Center [UNEP].)

and join GESAMP. Accordingly, the IOC Working Group on Marine Pollution was dissolved at the Sixth Session of IOC in Unesco, Paris, September 1969.

Other UN agencies accelerated their marine pollution studies, related to their responsibilities, during the 1970s. Some of the work was done through working groups in GESAMP. However, some projects were considered outside of the scope of GESAMP and were carried out unilaterally or jointly with other agencies having an interest in the marine environment. The Food and Agriculture Organization (FAO) has the responsibility of controlling and preventing marine pollution, as it affects living marine resources and fishing. The World Meteorological Organization (WMO) has a responsibility for the atmosphere, but has been quite active on sea-surface pollution problems, particularly if they arise from atmospheric input. The International Atomic Energy Agency (IAEA) has as its primary responsibility the coordination internationally of all matters pertaining to radioactivity. One aspect of this responsibility concerns radioactive waste disposal in the marine environment. The World Health Organization (WHO) is the international body with the responsibility for human health. Aside from taking a lead on matters related to human health in GESAMP, WHO has published a series of technical books on environmental health criteria, with respect to certain metals (mercury, cadmium and lead) and halogenated hydrocarbons (PCBs and PCTs), by organizing task groups to study the health implications of these substances. The United Nations itself has had an interest in certain types of marine pollution problems, particularly those related to coastal area development. This interest is expressed through the Ocean Economics and Technology Office of the Department of International Economic and Social Affairs. Through GESAMP, this body has been a lead agency in a Working Group on Sea-Bed Exploitation and Coastal Area Development.

The International Council for Exploration of the Sea (ICES), a regional non-UN body in the marine field since 1901, has been actively engaged in marine pollution work since the late 1960s. It has been a leader in baseline studies of the North Sea, in coordinating monitoring programs of the North Sea, and in conducting intercalibration exercises on metals analyses in sea water.

In the last five years (1975–1980), there have been comparatively few initiatives to develop new international conventions for marine pollution control. Most of the negotiations on this matter have taken place within the framework of the Law of the Sea Conference. In general, this has been a period of consolidation and implementation of the existing conventions. International programs of research and monitoring initiated in the first half of the 1970s were generally winding up during the second half of the decade and few new programs were replacing them. Nationally, in the western world, environmental authorities have been waging a constant struggle to keep existing environmental legislation from being weakened or overridden. This is already being reflected internationally with strong resistance to strengthening of existing conventions. Proposed amendments that would prohibit or better control the disposal of certain types of substances into the sea may meet with stiff opposition. During the 1980s, it is anticipated that other issues, such as energy and food supplies, will continue to replace environmental concerns, and the oceans could very well experience increased use for waste disposal as land disposal options for certain types of wastes become exhausted.

INTERNATIONAL CONVENTIONS ON PREVENTION OF MARINE POLLUTION

There are few global conventions for the prevention of marine pollution. A somewhat larger number of regional conventions exist. Table 1, starting with 1958, shows the important conventions, protocols, treaties and regulations established for both regional and global control of marine pollution. Where a state is signatory to both a regional and global convention, precedence will almost always be given to the former, inasmuch as there is overriding concern for that state to control pollution that directly affects it in the region.

The responsibility for enforcement of the provisions of a convention generally lie with the contracting parties. The organization that administers a convention usually has no enforcement authority. For example, the Inter-Governmental Maritime Consultative Organization is responsible for the Conventions for the Prevention of Pollution of the Sea by Oil, 1954, and for the Prevention of Marine Pollution by Dumping of Wastes and Other Matter, 1972, but its role is confined to a secretariat function in providing the necessary administrative, legal and communication services.

Conventions are usually developed by mutual agreement among participating states to institute a formal umbrella arrangement that will encompass forms of national legislation that exist or are proposed for marine pollution control. A convention is usually ratified by a state only after it has prepared national legislation that is at least as stringent as that provided for by the convention. For this reason, there is sometimes a long delay between the time a convention is developed and when it is brought into force.

Convention on the Prevention of Marine Pollution by Dumping of Wastes and Other Matter, 1972

This convention was formulated in a series of inter-governmental meetings during 1971 and 1972, when preparations were being made for the UN Conference on the Human Environment, and the national and international concerns about the marine environment were running high. There were a number of questionable episodes of dumping of wastes in the North Sea and the Northeast Atlantic, which stimulated action on the development of both a regional and global dumping convention. A Convention for the Prevention of Marine Pollution by Dumping from Ships and Aircraft, commonly known as the Oslo Convention (OC), was developed by western European countries during 1971 and 1972 and brought into force in 1974. Draft articles for a global dumping convention were first presented by the US delegation to the Second Session of the Intergovernmental Working Group on Marine Pollution, held in Ottawa, Canada, during November 1971. This was followed by meetings hosted by Iceland in Reykjavik, April 1972, and by the United Kingdom in London, May 1972, to review and revise the draft articles

TABLE 3.1

INTERNATIONAL CONVENTIONS AND AGREEMENTS FOR CONTROL OF MARINE POLLUTION

	POLLUTANT	RESPONSIBLE BODY	STATUS JUNE 1980
General Marine Pollution			
Convention on the Territorial Sea and Contiguous Zone	Various	UN	In force
Convention on the High Seas, 1958	Oil; wastes from exploration and exploitation of the seabed and its subsoil; and radioactive wastes	UN	In force
Convention on the Continental Shelf, 1958	Any harmful agents	UN	In force
Convention on Fishing and Conservation of the Living Resources of the High Seas, 1958	All deleterious substances	UN	Not yet in force
Antarctic Treaty	All deleterious substances	UN	In force
Convention on the Prohibition of the Development, Production and Stockpiling of Bacteriological (Biological) and Toxic Weapons and on their Destruction, 1972	Biologically hazardous and other toxic substances	UN	Not yet in force
Protocol Relating to Intervention on the High Seas in Cases of Marine Pollution by Substances Other Than Oil, 1973.	Other substances than oil	IMCO	Not yet in force
Convention on the Protection of the Marine Environment of the Baltic Sea (Helsinki Convention).	All substances from all sources	Helsinki Commission	Entered in force May 3, 1980
Convention for the Prevention of Marine Pollution from	Wastes from land based sources	French Government (Paris Commission)	In force

TABLE 3.1 *Cont.*

	POLLUTANT	RESPONSIBLE BODY	STATUS JUNE 1980
Land-Based Sources (Paris Convention) Convention on the Protection of the Environment between Denmark, Finland, Norway and Sweden	Environmentally harmful activities	Swedish Government	Not yet in force
Convention for the Protection of the Mediterranean Sea against Pollution (Barcelona Convention)	All environmentally harmful substances	UNEP	Adopted in February 1976. Entered into force February 12, 1978. Ratified by thirteen Mediterranean States and the EEC.
Protocol to the Barcelona Convention for Prevention of Pollution of the Mediterranean Sea by Dumping from Ships and Aircraft	All Substances	UNEP	Entered into force February 12, 1978.
Protocol to the Barcelona Convention for Co-operation in Combating Pollution in Cases of Emergency	Oil and other noxious substances	UNEP	Entered into force February 12, 1978.
Protocol to the Barcelona Convention for the Protection of the Mediterranean Sea against Pollution from Land-Based Sources	All environmentally harmful substances	UNEP	Expected to enter into forece in 1980.
Kuwait Regional Convention for Co-operation on the Protection of the Marine Environment from Pollution and the Protocol concerning Regional Cooperation in Combating Pollution by Oil and other Harmful Substances in Cases of Emergency	All harmful substances	UNEP (Kuwait Government)	Entered into force June 30, 1979. Regional Organization being established.

TABLE 3.1 *Cont.*

INTERNATIONAL CONVENTIONS AND AGREEMENTS FOR CONTROL OF MARINE POLLUTION

	POLLUTANT	RESPONSIBLE BODY	STATUS JUNE 1980
Oil Pollution			
International Convention for the Prevention of Pollution of the Sea by Oil, 1954.	Oil	U.K. until establishment of IMCO in 1959	Entered into force July 26, 1958.
Amendments to the International Convention for the Prevention of Pollution of the Sea by Oil, 1962.	Oil	IMCO	Entered into force May 18, and June 28, 1967.
Amendments to the International Convention for the Prevention of Pollution of the Sea by Oil, 1969	Oil	IMCO	Ratified in Jan. 19, 1977 by required number of Contracting Parties. Entered into force on Jan. 20, 1978.
(a) 1971 (Great Barrier Reef) Amendments	Oil	IMCO	Not yet in force.
(b) 1971 (Tanks) Amendments	Oil	IMCO	Not yet in force.
International Convention Relating to Intervention on the High Seas in Cases of Oil Pollution Casualties, 1969	Oil	IMCO	Entered into force May 6, 1975.
International Convention on Civil Liability for Oil Pollution Damage, 1969	Oil	IMCO	Entered into force June 19, 1975.
International Convention on the Establishment of an International Fund for Compensation for Oil Pollution Damage, 1971	Oil	IMCO	Entered into force October 16, 1978.
International Convention for the Prevention of Pollution from Ships, 1973 (MARPOL Convention)	Oil and other substances carried and/or discharged by ships	IMCO	Not yet in force
Protocol of 1978 relating to the International Convention	Oil pollution by ships	IMCO	Not yet in force

TABLE 3.1 *Cont.*

INTERNATIONAL CONVENTIONS AND AGREEMENTS FOR CONTROL OF MARINE POLLUTION

	POLLUTANT	RESPONSIBLE BODY	STATUS JUNE 1980
for the Prevention of Pollution from ships, 1973			
Agreement Concerning Cooperation in Dealing with Pollution of the North Sea by Oil, 1969	Oil	Government of the Federal Republic of Germany	In force
Agreement Concerning Cooperation in Measures to Deal with Pollution of the Sea by Oil, 1971	Oil	Danish Government	In force
Radioactivity			
Convention on Third Party Liability in the Field of Nuclear Energy, 1960	Radioactive materials	UN, IAEA	In force
Convention Supplementary to the Paris Convention on Third Party Liability in the Field of Nuclear Energy, 1963	Radioactive materials	UN, IAEA	Not yet in force
Convention on the Liability of Operators of Nuclear Ships, 1962	Radioactive materials	UN, IAEA, IMCO	Not yet in force
Convention on Civil Liability for Nuclear Damage, 1963	Radioactive materials	UN, IAEA	Not yet in force
Treaty Banning Nuclear Weapons Tests in the Atmosphere, in Outer Space and Underwater, 1963	Radioactive materials	UN, IAEA	In force
Treaty on the Prohibition of the Emplacement of Nuclear Weapons, and other Weapons of Mass Destruction on the Sea-Bed and Ocean Floor and on the Subsoil Thereof, 1971	Radioactive materials	UN, IAEA	In force
Convention Relative to Civil Liability in the Field of Maritime Carriage of	Radioactive materials	IMCO, IAEA	Entered into force July 1975.

TABLE 3.1 *Cont.*

INTERNATIONAL CONVENTIONS AND AGREEMENTS FOR CONTROL OF MARINE POLLUTION

	POLLUTANT	RESPONSIBLE BODY	STATUS JUNE 1980
Nuclear Material, 1971			
Regulations for the Safe Transport of Radioactive Materials	Radioactive materials	UN, IAEA	Adopted
Regulations for the Safe Transport of Radioactive Materials	Radioactive materials	UN, IAEA	Adopted
Basic Safety Standards for Radiation Protection	Radioactive materials	UN, IAEA	Adopted
Standardization of Radioactive Waste Categories	Radioactive materials	UN, IAEA	Adopted
Regulations for the Safe Transport of Radioactive Materials	Radioactive materials	UN, IAEA	Adopted
Ocean Dumping			
Convention for the Prevention of Marine Pollution by Dumping from Ships and Aircraft (Oslo Convention)	All wastes and other substances dumped at sea	Oslo Commission	In force.
Convention for the Prevention of Marine Pollution by Dumping of Wastes and Other Matter (London Dumping Convention)	All wastes and other matter dumped at sea	IMCO	Entered into force August 30, 1975.
(a) 1978 Amendments on Procedures for the Settlement of Disputes	All wastes and other matter	IMCO	Not yet in force.
(b) 1978 Amendments on the Prevention and Control of Pollution by Incineration of Wastes and Other Matter	All wastes and other matter	IMCO	Entered into force March 11, 1979, except for Federal Republic of Germany and New Zealand.

SOURCE: Modified from Waldichuk, 1978

for a global ocean dumping convention. The conference to formally develop and sign the convention was attended by delegates from about seventy-five countries in London during October and November 1972. Ratification by the necessary fifteen states took place quite rapidly, and the Convention on the Prevention of Marine Pollution by Dumping of Wastes and Other Matter, 1972, commonly known as the London Dumping Convention (LDC), was brought into force in 1975. Many of the states signatory to the Oslo Convention also ratified the London Dumping Convention.

The London Dumping Convention was developed on the basic premise that marine pollution by dumping of wastes and other matter should be prevented by prohibiting such dumping with certain exceptions. In fact, as given in annex 1 of the convention, only certain substances (organo-halogen compounds, most petroleum hydrocarbons, mercury and its compounds, cadmium and its compounds, persistent plastics, high-level radioactive wastes, biological and chemical warfare agents) are actually prohibited, but not if they are present only in trace amounts and not if they are rapidly rendered harmless by physical, chemical or biological processes in the sea. Annex 2 of the convention lists substances that require special care in ocean dumping and includes: a. arsenic, lead, copper, zinc and their compounds, organosilicon compounds, cyanides, fluorides, pesticides and their by-products not covered in annex 1; b. beryllium, chromium, nickel, vanadium and their compounds if contained in acidic or alkaline effluents; c. containers, scrap metal and other bulky solid wastes which may be an obstacle to navigation and fishing; and d. radioactive materials not listed in annex 1. Substances in category a require special care and the issuance of a special permit only if they are contained in wastes in significant amounts. All other substances dumped into the sea require only a general permit. There are other exclusions for annex 1 substances, as contained in article 5 on emergency dumping. Careful consideration has to be given to characteristics of the material to be dumped and of the dumpsite. The important factors in this connection are set out in annex 3 of the convention.

In accordance with article 14, a competent organization to administer the convention should be designated by contracting parties within three months of entry into force of the convention. IMCO was designated as this organization. Consultative meetings of contracting parties to the convention are arranged annually, and the fifth consultative meeting is scheduled for September 1980. Ad hoc groups have been set up on legal matters, scientific aspects and incineration at sea under the convention. One legal matter that required early attention, as specified under article 11, was the development of procedures for the settlement of disputes concerning the interpretation and application of the convention. In spite of objections of some countries that this should await the successful conclusion of the United Nations Conference on the Law of the Sea, procedures were adopted by a majority vote at the third consultative meeting in October 1978.

The main function of the Ad Hoc Group on Incineration at Sea was to develop technical guidelines on the control of incineration of wastes and other matter at sea, and procedures for the implementation of paragraphs 8 and 9 of annex 1, dealing with substances rapidly rendered harmless and materials containing annex 1 substances as trace contaminants. With the aid of the legal group, these guidelines and procedures have been appropriately incorporated into the convention. One major objection on incineration at sea comes from the Scandinavian countries, which would like to see such practice discontinued as early as possible and other means developed to destroy dangerous substances, e.g. PCBs, DDT and other organohalogens.

The Ad Hoc Scientific Group on Dumping has taken on itself the task of preparing amendments for annexes 1–3, and to define such terms given in the annexes as trace amounts, rapidly rendered harmless, and significant amounts. Although provision for amendments to the convention and its annexes is clearly spelled out in article 15, and early amendment of the annexes was generally accepted by participants at the conference that developed the convention, there has been strong resistance to adopting proposed amendments, even though these might be fully supported by available scientific and technical information. Part of this resistance is due to legal and administrative constraints in the amendment of national legislation of contracting parties to harmonize it with the amended convention. The other reason for strong resistance to amendments is the effect this would have on other conventions, none the least of which is the Oslo Convention. As an example of this, the proposal to include lead and lead compounds in annex 1 has met strong opposition from some states, not on a clear scientific basis, but on administrative grounds in that it could have an impact on the Paris Convention, and regulations associated with it, concerning substances in land-based sources of pollution.

A proposal to amend annex 2 with an item "Oxygen consuming and/or biodegradable organic matter" met with much opposition in the ad hoc scientific group until it was pointed out to be similar to a paragraph in annex 2 of the Oslo Dumping Convention: "1. (c) Substances which, though of a non-toxic nature, may become harmful due to the quantities in which they are dumped, or which are liable to seriously reduce amenities." Thereupon, a delegate from one of the contracting parties to the Oslo Dumping Convention stated that if the proposed amendment to annex 2 of the LDC were put in exactly the same words as in the ODC, it would be acceptable, even though the latter was in effect requiring a higher degree of control than in the original proposed amendment, in that it considers control to prevent impairment of amenities.

The same problem applies to definition of identical or similar terms in the two conventions. The presence of annex 1 substances as trace contaminants in the LDC was defined by a set of guidelines only after similar procedures had been developed to establish a trace contaminant in the ODC. In spite of many attempts to define a "significant amount" of annex 2 material, none has been accepted, and the interim definition of "0.1% by weight" continues to be used for both conventions.

It is clear that conventions should be formulated quite precisely when they are developed, because it takes a long time to change them once they are in force, even though amendments might effect substantial improvement. Certainly there is a dominant influence of the Oslo Convention on any substantive changes that contracting parties may want to introduce into the London Dumping Convention. But the inertia to change in the London Dumping Convention, especially if amendments make the convention more stringent, may also be a manifestation of relaxation in national legislation on marine pollution control.

International Convention for the Prevention of Pollution from Ships (MARPOL), 1973

This convention was developed in response to a need for major amendment of the International Convention for the Prevention of Pollution of the Sea by Oil, 1954, as amended in 1962 and 1969. IMCO undertook during 1970–73 to canvass member states on the problems of marine pollution from ships. Through the Subcommittee on Marine Pollution of the IMCO Maritime Safety Committee, a draft convention was prepared for a diplomatic conference held in London in October 1973. Basically, three sources of marine pollution associated with ships were considered: (1) oil transported by ships; (2) hazardous substances other than oil transported by ships; and (3) ship-generated wastes (sewage and garbage). The convention, as it was developed by the end of the conference in 1973, was signed by the seventy-five participating states. Uniform control over pollution from ships would be provided globally, but three special areas were designated to receive high-level protection: the Mediterranean, Baltic, and Black Seas.

The convention will enter into force a year after it has been ratified by fifteen states with at least 50% of the world's gross tonnage of merchant shipping. As of December 1980, nine contracting states to the convention (United Kingdom, Norway, Yugoslavia, Jordon, Kenya, Tunisia, Yemen, Uruguay and Peru) have acceded to it. It could be a long time before the required number of states with the necessary tonnage ratify the convention in order to bring it into force. For this reason, a protocol was developed in 1978 to accelerate acceptance of the vital components of the convention dealing with prevention of oil pollution.

The 1978 Protocol to the 1973 MARPOL Convention incorporates those sections of the parent convention relating to oil pollution, and excludes others, in an attempt to bring the oil-pollution control measures into force as soon as possible. The more technically difficult annexes of the convention, particularly those dealing with pollution from chemical substances, could be deferred. So far, the protocol has been ratified by one state, one country has started ratification procedures, ratification is under preparation in four states and under consideration in five others. The protocol is expected to be ratified by nine countries in 1980. The same level of acceptance is required by the protocol as by the parent convention before it can be brought into force. In the meantime, the Marine Environment Protection Committee (MEPC) of the Inter-Governmental Maritime Consultative Organization continues to meet (the fifteenth session was held in April 1980) to provide interpretation and to develop implementation procedures for the convention and the protocol. Member states are endeavouring to bring their national legislation, dealing with pollution from ships, into line with the MARPOL Convention, but there appear to be many stumbling blocks. The matter of prosecution in the event of an infraction of the MARPOL Convention in international waters is left to the flag state, and this may be a weakness in the convention.

Other International Conventions

Many of the other international conventions, as shown in table 1, deal with regional problems, where a number of countries border on a particular sea area, such as the Baltic or Mediterranean Sea. It is expected that conventions for other marginal seas, e.g. Red Sea/Gulf of Aden and Caribbean, will be developed under UNEP's Regional Seas Programme.

Conventions on marine pollution administered by IMCO, in addition to the two already noted, deal largely with Civil Liability for Oil Pollution Damage, Establishment of an International Fund for Compensation for Oil Pollution Damage, and Intervention on the High Seas in Cases of Oil Pollution Casualties (table 1). Because of the responsibility of IMCO to maritime shipping, most of the conventions under its wing are ship-oriented. At the forty-first session of the IMCO Legal Committee, a proposed convention was discussed on liability and compensation in connection with the carriage of noxious and hazardous substances by sea. It is planned by IMCO to convene a diplomatic conference in 1982, at which time it is hoped that a new convention covering the foregoing issues will be adopted.

Any future conventions on the prevention and control of marine pollution will have to be rather narrowly defined in terms of responsibility, to ensure that they are brought into force reasonably soon after their adoption and can be effectively administered. The declaration of 200-mile economic zones by many maritime countries of the world has increased national jurisdiction and reduced the area over which international conventions can be generated. Conventions arising out of the Third UN Conference on the Law of the Sea are expected to cover areas beyond the limits of national jurisdiction. The International Sea-Bed Authority, if it becomes a reality under the New Law of the Sea, will probably be given far-reaching powers to control pollution from exploration and exploitation of the sea-bed for minerals in international waters, e.g. ferro-manganese nodules in the eastern equatorial Pacific. Except in marine areas of adjoining state jurisdictions, it is not anticipated that landbased sources of marine pollution will be controlled under international conventions. Sovereign rights in territorial seas and 200-mile economic zones are rather jealously protected by all states.

SOURCES OF MARINE POLLUTION STILL REQUIRING CONTROL

As noted earlier, marine pollutants recognize no boundaries. It makes little difference over the long term where a pollutant is released into the sea; theoretically, it can eventually reach any part of the world oceans, depending on its persistence. Hence, land-based sources of pollution can be as important to the global marine pollution picture as deep-sea sources beyond the limits of national jurisdiction. However, the discharge of substances within the limits of national jurisdiction is generally considered to be a national responsibility. Any attempt by another state or by an international body to control such discharges is usually regarded as interference into national affairs. It is only when compelling evidence is produced to demonstrate that pollutant discharges of a given state or states are causing damage to the waters of another coastal state or states, or are leading to a dangerous trend in international waters, that positive action can be generated toward an international agreement. This kind of evidence is usually difficult to produce, except in those extreme situations where the health of humans on a world-wide scale is shown to be threatened, as appears to have occurred prior to the signing in 1963 of the Treaty Banning Nuclear Weapons Tests in the Atmosphere, in Outer Space and Under Water.

Before we can safely say that pollution of the global oceans is fully under control, we must obtain a good inventory of the input from the numerous sources. We must be able to assign a relative importance, in terms of contribution to a particular pollutant load, from each of the sources. Then an effort must be made to control the most important sources of pollution. There is a tendency to regulate those sources of pollution that yield to easy control and to hold in abeyance those sources that are difficult to manage. This does not always mean that the most important pollutant sources are controlled.

As pollutant sources to the oceans, ships are perhaps the most amenable to regulation. Except for oil pollution from tankers and waste discharges from vessels in busy coastal seaways and harbours, however, it is doubtful whether shipping is a significant contributor to most of the critical marine pollutants. Atmospheric transport from land to sea has been considered as an important route for such marine pollutants as DDT and other chlorinated hydrocarbons. Yet the non-point sources of these materials to the atmosphere, e.g. agricultural and forest spraying, are difficult, if not impossible, to control. Ocean outfalls can be monitored and controlled, but most of them are still rather loosely managed. River-borne pollutants may be major contributors, but even if an accurate accounting is available of materials introduced into rivers, we have little information on the proportion of the river-borne material that actually escapes to sea through the estuaries.

The exploitation of the sea-bed for petroleum and hard minerals on the continental shelf and slope is not controlled internationally at the present time and probably never will be, because the area generally comes under national jurisdiction. The degree of national control of this pollution source varies from country to country, and in some cases it may be totally uncontrolled. Pollution from exploitation of sea-bed minerals in the deep sea beyond the limits of national jurisdiction will probably be controlled by the International Sea-Bed Authority under the new Law of the Sea. National authorities as well as international agencies are looking foreward to see how this experiment in international management of a resource, termed as "the common heritage of mankind", will work.

The marine pollutant sources which are still largely uncontrolled internationally are given in Table 2.

INTERNATIONAL SCIENTIFIC AND TECHNICAL ACTIVITIES ON MARINE POLLUTION

The objectives of international scientific and technical activities on marine pollution have been mainly to: (1) review available information to

TABLE 3.2

ASPECTS OF MARINE POLLUTION REQUIRING AGREEMENTS FOR INTERNATIONAL CONTROL

	NATIONAL CONTROL	INTERNATIONAL CONTROL	COMMENTS
Discharges from ships	Develop national legislation and regulations for the prevention and control of pollution in coastal waters (territorial sea, economic zone, fishing zone, and in special areas within those zones).	Establish a sound basis of regulations for the prevention and control of pollution from vessels in the territorial sea, in the economic zone and in special areas within the zone, as well as in international waters.	Must be harmonized with the International Convention for the Prevention of Pollution from Ships, 1973
Sea-bed mining from drill rigs, drill ships and artificial islands	Activities in territorial sea, economic or fishing zone come under national control.	Activities beyond the limits of national jurisdiction could be controlled by an International Sea-Bed Authority.	New set of guidelines and regulations must be developed for control internationally.
Discharge from ocean outfalls	Virtually all such discharges will be under national control. Coastal state standards must be uniformly based on international guidelines.	Criteria, standards and regulations must be developed to bring all effluent discharges under uniform control for marine environmental preservation.	Most such discharges are within national boundaries of the coastal zone. Helsinki and Paris Conventions and Protocol to the Barcelona Convention consider such land-based sources.
Atmospheric emissions	All such emissions come under control of local authorities within the state. Controls could be based on international guidelines.	Criteria, standards and regulations for atmospheric emissions must be developed to provide uniformity for protection of the atmosphere and aquatic resources.	Certain atmospheric emissions, e.g. sulphur dioxide, may be controlled for protection of freshwater systems.
Other land-based sources	Coastal and inland states contributing to coastal pollution through runoff, rivers, spills, and leakage can control these sources through legislation and technological developments for clean up.	Development of effluent guidelines, water quality criteria and treatment technology handbooks for use by national authorities in marine pollution control is a first step.	A very difficult problem to resolve internationally, because of possible interference in national affairs. Paris Convention provides precedent.

SOURCE: Waldichuk, 1978.

identify the problems and to recommend solutions; (2) assist in the development and/or administration of conventions on prevention of marine pollution; and (3) assist developing countries through technical assistance, training by courses and workshops, and involvement in regional marine pollution investigation.

Specialized international bodies, such as the International Atomic Energy Agency (IAEA), regularly examine the state of knowledge on marine pollution related to their particular specialty by convening symposia, workshops and panels and subsequently publishing the proceedings, which may become the definitive statements on the particular aspects covered. In the case of IAEA, their International Laboratory on Marine Radioactivity in Monaco regularly makes significant contributions, through laboratory and field research, to knowledge on radioactivity in the marine environment.

Through such bodies as GESAMP, international agencies with different mandates have been able to deal with marine pollution problems in an interdisciplinary way. The radioactivity problem may be examined in terms of dispersion in the marine environment, bioaccumulation, and effects on human health. Thus marine biologists are able to interact with marine chemists, physical oceanographers, meteorologists, public health specialists and sanitary engineers. This kind of dialogue provides an insight into some of the marine problems that would not be possible from the point of view of one particular discipline.

The details of various studies conducted by international working groups, panels and task teams from 1965 to 1977 have been reported elsewhere (Waldichuk 1973, 1978). Some of the more recent and current projects will be reviewed here. Many of them are oriented to developing countries.

Technical Assistance by IMCO with Support from UNEP

Regional Oil Combating Centre for the Mediterranean Sea, Malta

This center has been operated since 1976 by IMCO with support of UNEP. It was established as part of the drive to clean up the Mediterranean spearheaded by the United Nations Environment Programme. Its purpose is to develop contingency plans for combating oil pollution and to assist Mediterranean countries in the event of an oil spill, particularly from tankers. There have been recent recommendations to strengthen the center for assisting with development of sub-regional contingency plans and to provide on-the-spot technical assistance in cases of marine pollution emergencies.

Marine Emergency Mutual Aid Centre (MEMAC)

Part of the Kuwait Action Plan, this center has been recommended for establishment in Bahrain, for the purpose of information exchange, technological cooperation and training and possible future extension of responsibilities to initiate operations on marine pollution problems. Preparatory work for its establishment has been completed.

Overview Study of Oil Pollution in the Wider Caribbean Region

This "desk study" was initiated by IMCO, in support of UNEP's Action Plan for the Wider Caribbean Area, to provide an assessment of the extent of oil pollution in the region, together with a description of measures taken by governments and industry to prevent and mitigate the effects of such pollution.

Development of Contingency Plans for Smaller Island Countries of the Caribbean Region

Undertaken by IMCO and UNEP jointly with the Organization of American States (OAS) and with assistance from the U.S. Agency for International Development (USAID), this project was a first step in the development of a Caribbean Oil Spill Program. It will involve preparation of an action plan for controlling pollution for oil spills and a training program in 1981.

Research into Toxicity of Oil Dispersant Chemicals on Tropical and Sub-Tropical Marine Species

A program of toxicity testing on oil dispersant chemicals has been initiated in Philippine laboratories, with the aid of a consultant experienced in bioassays, to determine the suitability of oil dispersing chemicals for use in tropical and sub-tropical waters.

Survey of Oil Pollution of the West African Coast with Particular Emphasis on Pollution from Shipping Activities

Designed to assist governments of the region in identifying priority actions for the control of oil pollution, as part of UNEP's Action Plan for West Africa, this project provides an overview study of the present state of marine pollution by oil in the area, with particular emphasis on maritime activities. The report covering this study will be used as a background document for the plenipotentiary conference that UNEP plans to convene in the region early in 1981.

Consultancy Projects

With the aid of funds from the United Nations Development Programme (UNDP), several consultancy projects were completed during 1978 and 79: Consultancy on Marine Pollution from Ships – Jamaica and Management Support to the National Port Authority of Panama (Consultancy in Pollution Control). When the new Panama Canal Treaty came into force on October 1, 1979, the National Port Authority of Panama took over the responsibility from the Panama Canal Company for the management of the ports of Balboa and Cristobal. The project coordinated by UNCTAD (United Nations Conference on Trade and Development) in association with ILO (International Labour Organization) and IMCO was intended to ensure a smooth transition.

Prevention of Marine Pollution Caused by Oil – China

Developed in two parts and funded by UNDP, this project was designed to assist China's efforts to protect its seas from oil pollution. In the first stage, a two-month study tour during April and May 1980 was conducted by Chinese engineers and scientists visiting centers in Japan, North America and Europe to study methods of preventing and dealing with oil pollution. The second stage will be initiated in late 1980 when Chinese specialists will attend intensive anti-pollution courses in foreign centers.

Technical Symposium on Marine Pollution and Study Tour – China

Scheduled for late September 1980 with funding by UNDP, the one-week symposium will provide a forum for technologists from Asian countries to discuss requirements of IMCO pollution prevention conventions. Emphasis will be placed on shipboard equipment, such as oily-water separators and oil content meters, and their performance standards, as well as port reception facilities. A one-week study tour of Chinese ports will follow the symposium.

Prevention and Control of Offshore Oil Spills – India

This project, formulated and funded by UNDP to cover a two-year period 1980–81, involves IMCO in providing emphasis on oil pollution preventive measures, and is concerned primarily with the Bombay High

Oil Field. Training of local personnel in pollution preventive measures is one of the main responsibilities of IMCO.

Combating of Marine Pollution by Oil – Chile

Another UNDP-financed project for a two-year period starting April 1980, the objective is to implement a national contingency plan by training oil pollution control officers in the United States. In-service training of other personnel, provision of equipment (skimmers, booms, dispersant spraying equipment, dispersants and communications equipment) and development of anti-pollution courses which will be available to candidates from other countries.

Freshwater Ballast Project

This project examined the feasibility of transporting fresh water instead of sea water as ballast in tankers returning to oil-rich but freshwater-poor countries of North Africa. The study, conducted by IMCO consultants for the Arab Development Institute, Tripoli, Libya, demonstrated that transport of fresh water for irrigation purposes by oil tankers returning to Libya would be economically feasible. Some problems could arise if the fresh water were contaminated by heavy metals or sewage pathogens.

Seminars and Courses

SIDA (Swedish International Development Agency) financially supported the following: the Third Course for Maritime Administrators on the Prevention and Control of Marine Pollution from Ships in Malmö, Sweden, during August 1979; a Latin American Seminar on Tanker Safety and Pollution Prevention in Rio de Janeiro, Brazil, in January 1980; and an African Regional Seminar on Tanker Safety and Pollution Prevention planned for Nairobi, Kenya, in early 1981. SIDA has also supported production of a film and a set of slides by IMCO on the 1973 MARPOL Convention, as modified by the 1978 protocol, for educational and training purposes.

Inter-regional Advisors and Consultants

The IMCO marine pollution adviser, based in London, had discussions during 1979 in Gabon, Cape Verde and Senegal. Two IMCO regional marine pollution advisers are based in Latin America. The inter-regional consultant on marine pollution undertook a mission to six of the island countries in the Caribbean in connection with the development of sub-regional contingency plans; he also visited, during 1979, Malaysia, Mauritius, Democratic Yemen, Seychelles, Mexico, Papua New Guinea, Fiji, Singapore and Sri Lanka.

UNEP Coordination of Activities in the Regional Seas Program

Mediterranean

The MED/POL program has been underway since 1975 and now has eighty-four laboratories involved from sixteen states making a scientific assessment of the Mediterranean's pollution problems. The mariculture and energy projects have made notable progress.

Red Sea

Following adoption of the action plan for the Red Sea and the Gulf of Aden at the Second Jeddah Conference on the Protection of the Marine Environment in the Red Sea Area in January 1976, UNEP decided to implement the Red Sea program through the Arab League Educational, Cultural and Scientific Organization (ALECSO). IMCO undertook coastal pollution surveys, under funds-in-trust arrangements with ALECSO, of Jordan's coastline in the Gulf of Aqaba and Egypt's coast-line in the Gulf of Suez and the Red Sea. Otherwise, activities have concentrated since 1976 on training and technical support to improve the capabilities of scientists and institutions in the region. A Regional Marine Mutual Aid Centre is planned. A number of seminars, work-shops, study tours and training courses have already taken place; a symposium on the coastal and marine environment of the Red Sea, Gulf of Aden and tropical western Indian Ocean is planned for Khartoum under UNESCO's cooperative arrangements in 1980.

Kuwait Action Plan Region.

The Kuwait Regional Convention for Co-operation on the Protection of the Marine Environment from Pollution and the Protocol Concerning Regional Co-operation in Combating Pollution by Oil and Other Harm-ful Substances in Cases of Emergency entered into force in June 1979. Priority action, as provided for in article 16 of the convention and article 3 of the protocol, required the establishment of the Regional Organi-zation for the Protection of the Marine Environment and the Marine Emergency Mutual Aid Centre (MEMAC), respectively. Priority co-operative projects identified in the action plan are being prepared for operational initiation. An International Workshop on Combating of Marine Pollution from Ships and Safety of Navigation, as part of the action plan, has been proposed by IMCO.

Caribbean

A draft action plan was prepared at a UNEP meeting of government-nominated experts to review the preliminary draft action plan for the Wider Caribbean Region, in Caracas, January 1980. The UNEP/ECLA Caribbean Environment Project, in cooperation with specialized agencies, is developing project proposals to be implemented under the Caribbean action plan.

West Africa

A meeting of experts was convened by UNEP in Libreville, Gabon, during November 1979 to review the Draft Action Plan for the Protection and Development of the Marine Environment and Coastal Areas of the West African region. The UN Department of International and Social Affairs recently organized a workshop on coastal erosion in Togo, which contributed to the action plan being developed for West Africa. In preparation for the intergovernmental meeting at the plenipotentiary level to review the draft action plan, IMCO has undertaken for UNEP a "Survey of Oil Pollution of the West African Region Coast with Particular Emphasis on Pollution from Shipping Activities".

East Asian Sea

A meeting of experts was convened in the Philippines, in June 1980, to review the draft action plan for the East Asian Seas. In preparation for the Meeting, UNEP initiated a project with support to IMCO and the Indonesian Government to jointly organize a meeting on the Development of Sub-Regional Oil Spill Contingency Arrangements in the Celebes (Sulawesi) Sea, which was held in Jakarta in January 1980. A risk analysis and an action plan for the Celebes Sea was agreed to by the meeting. A draft Celebes (Sulawesi) Sea Oil Spill Response Plan was appended to the action plan, which will eventually form the basis for a coordinated anti-pollution response capability in the sub-region. IMCO is planning an International Workshop on the Prevention, Abatement and Combating of Pollution from Ships in South-east Asian Waters, tentatively scheduled for Manila in 1980.

South-East Pacific

A meeting of regional experts was held in Santiago, Chile, in November 1978 to develop a framework for an Action Plan for the Protection of the Marine Environment in the South-East Pacific. The action plan is being developed jointly by UNEP with the Permanent Commission for the South Pacific (CPPS) based in Lima, Peru. IMCO has been invited to organize an international workshop for the prevention, combating and abatement of pollution for ships in the South-East Pacific in the latter part of 1980. With co-operation of FAO, IMCO and IOC, a program on marine pollution monitoring and research is being developed. Legal experts are formulating a regional convention on the protection of the marine environment against pollution, and an agreement on regional co-operation for emergency measures in the event of a major pollution incident.

South-West Pacific

A regional conference on the human environment in the South-West Pacific is planned for the second half of 1980, or soon thereafter under the joint sponsorship of the South Pacific Commission (SPC), the South Pacific Bureau for Economic Cooperation (SPEC), the Economic and Social Commission for Asia and the Pacific (ESCAP) and UNEP. It is expected to lay the groundwork for future environmental action at all levels of government in the region. A report on the state of the environment in the South-West Pacific will be submitted to the meeting along with a draft regional action plan. IMCO's participation in UNEP's Regional Seas Programme for the area will probably take the form of a study of co-operative arrangements in the region for combating pollution arising from emergencies.

Initiatives of the Intergovernmental Oceanographic Commission (IOC)

IOC has been involved in a number of marine projects where other agencies, such as FAO, WMO and UNEP, have taken the lead. There are several, however, in which IOC has been the leading agency. In general, it assumes the scientific role, with an element of training, education and mutual assistance, in studies of the marine environment. Workshops, seminars and training courses are used to improve the capabilities of technical personnel and facilities in developing countries through the IOC Working Committee on TEMA (Training, Education and Mutual Assistance in the marine sciences).

In the development of the Scope of the Long-Term and Expanded

Program of Oceanic Exploration and Research (LEPOR) at a meeting of a special working group of the IOC in Paris, June 1969, it was recognized that studies of marine pollution should be a component of the program, and a number of projects were recommended, including laboratory studies on delayed and sublethal effects of pollutants and establishment of a world-wide system of monitoring marine pollution constituents. LEPOR was generally accepted at the sixth session of IOC in September 1969, and instructions were given that the component parts of LEPOR be appropriately developed. This led to a recommendation of an IOC Group of Experts on Long-Term Scientific Policy and Planning (GELTSPAP) in Monaco, November 1970, that a Global Investigation of Pollution in the Marine Environment (GIPME) be a focal point of LEPOR. Subsequent actions by the IOC Bureau and Consultative Council and by the IOC Executive Council ensured that GIPME would be supported.

The International Decade of Ocean Exploration (IDOE), mainly a U.S. initiative, was also accepted at the sixth session of IOC, as an element of LEPOR. Within IDOE, a number of valuable marine pollution research projects have been developed, including the Controlled Ecosystem Pollution Experiment (CEPEX), Pollutant Transfer Program, Sea/Air Chemical Exchange (SEAREX), and Pollutant Responses in Marine Animals (PRIMA), all funded by the National Science Foundation Office for the International Decade of Ocean Exploration.

Global Investigation of Pollution in the Marine Environment (GIPME)

The rationale for GIPME was to provide expertise to member states in the field of marine pollution research and monitoring. At the first session of the International Coordination Group (ICG) for GIPME, a recommendation arose for preparation of a document on the health of the ocean, which was eventually published (Goldberg 1976). This is now being updated by a GESAMP working group on a review of the state of health of the oceans.

At the second and third sessions of the ICG for GIPME, a comprehensive plan for GIPME and baseline study guidelines were prepared (IOC 1976). The significant achievement of the first three sessions of the working committee of GIPME was the formation of the Group of Experts on Methods, Standards and Intercalibration (GEMSI), which undertook as its first task intercalibration exercises under the IOC/WMO/UNEP Pilot Project on Monitoring Background Levels of Selected Pollutants in Open Ocean Waters. With chlorinated hydrocarbon analytical standards prepared by IAEA's International Laboratory of Marine Radioactivity in Monaco and analytical standards for some trace metals prepared by the Sagami Chemical Research Centre in Japan, GEMSI organized an intercalibration for both organochlorines and metals during 1980 involving about 100 participants. Part of the exercise involved intercalibration of sampling from the open ocean in the Bermuda area in January 1980 within the opening phase of the Pilot Project on Monitoring Background Levels of Selected Pollutants in Open Ocean Waters. The Bermuda experiment was carried out as planned, with IOC sponsorship and international participation, Canada taking the lead role on metals intercalibration and Norway on petroleum hydrocarbons and chlorinated hydrocarbons.

The IGOSS Marine Pollution (Petroleum) Pilot Project (MAPMOPP)

The project commenced in January 1975 within IOC's Integrated Global Ocean Station System (IGOSS), using ships of opportunity to sample waters of the North Atlantic to determine the amount of oil (including tar balls, films and dissolved oil) present in surface waters. Approved initially for a period of two years, it was extended to the end of 1979 with the future of the project to be determined by results of a review of the data by an ad hoc group of experts on the evaluation of MAPMOPP meeting in Tokyo in July 1979 and the Third IOC/WMO Workshop on Marine Pollution (Petroleum) Monitoring in New Delhi, in February 1980. The results so far, as least for visual observations and for floating tar balls, indicate that the program has been a success, and that the sort of data acquired can be handled by existing data systems. The interpretation however, of UV-spectro-fluorometric data on dissolved/dispersed hydrocarbons has been questioned. It is uncertain also whether the system could be used successfully for monitoring other pollutants. It was proposed, nevertheless, that MAPMOPP be phased into a more permanent monitoring activity under MARPOLMON (Marine Pollution Monitoring) during mid-1980, with further evaluation taking place at the fourth session of the working committee on GIPME in 1981.

Mediterranean Pollution Study Pilot Projects (MEDPOL):
MED-I. The IOC/WMO/UNEP Pilot Project on Baseline Studies and Monitoring of Oil and Petroleum Hydrocarbons in Marine Waters. MED-VI. The IOC/UNEP Pilot Project on Problems of Coastal Transport of Pollutants.

In both these projects, IOC has played a leading role, with MED-I being essentially an extension of MAPMOPP.

IOCARIBE/WECAFC Marine Pollution Program

Framed at the second session of IOCARIBE (San Jose, Costa Rica, August 1978) after full intersecretariat discussion with WECAFC, this program is expected to continue until full co-ordination with the Caribbean Environment Programme is feasible.

An IOC/FAO/UNEP International Workshop on Marine Pollution in the Caribbean and Adjacent Regions was held in Port-of-Spain, Trinidad and Tobago, December 1976, and efforts are being made to follow up on its recommendations.

IOC/FAO/WHO/UNEP International Workshop on Marine Pollution in the Gulf of Guinea and Adjacent Areas, Abidjan, Ivory Coast, May 1978

Organized primarily to identify marine pollution problems in the area and how these might be investigated, the workshop led to development by UNEP of a draft action plan. GEMSI has been instructed by the working committee for GIPME to review the report of the workshop and eventually scrutinize programs resulting from its recommendations with a view to providing technical advice to IOC member states on their implementation.

CPPS/FAO/IOC/UNEP International Workshop on Marine Pollution in the Southeast Pacific. Santiago, Chile, November 1978

This workshop was organized at the request of the Permanent Commission for the South Pacific (CPPS) as an initial step to provide training and technical assistance in order to increase the capabilities of the laboratories in the region.

Scientific and Technical Activities of other UN Agencies on Marine Pollution

World Health Organization (WHO)

WHO is involved in various marine studies related to its mandate to protect human health. It has taken a leading role in connection with the study of the effects of sewage on human health through contamination of beaches used for bathing and of shellfish used for human consumption. Some studies have been done cooperatively with FAO on effects of pollution on the quality of seafood, and some have actually examined sub-lethal effects on marine organisms (WHO, 1975). WHO has been involved in studying the human health implications of nuclear power (WHO, 1978), in addition to studies by IAEA, and environmental health impact assessment (WHO, 1979).

International Atomic Energy Agency (IAEA)

IAEA has been a source of authoritative information through conferences, symposia and panel reports on radioactivity in the marine environment and concerning problems of radioactive waste disposal into the sea. It is consulted on radioactive matters when international conventions and treaties are formulated, or when interpretations or definitions are required, e.g., the definition of high-level radioactive wastes in annex 1 of the Convention for the Prevention of Marine Pollution by Dumping of Wastes and Other Matter.

World Meteorological Organization (WMO)

WMO has responsibility for the atmosphere and for substances and processes therein. It has also had an interest in the air/sea interface and on the transfer of substances and heat through this interface. In that connection, it has worked closely with the IOC on matters related to the air/sea boundary and has always played an important role in IGOSS. WHO has been the lead agency in the GESAMP Working Group on Interchange of Pollutants between the Atmosphere and the Oceans.

Food and Agriculture Organization of the United Nations (FAO)

FAO has a mandate for protection against pollution of living resources in the sea. It organized and convened the FAO Technical Conference on Marine Pollution and its Effects on Living Resources and Fishing in December 1970. The papers presented at this conference (RUIVO, 1972) provide a significant contribution to the marine pollution literature. The Seminar on Methods of Detection, Measurement and Monitoring of Pollutants in the Marine Environment (FAO 1971), which preceded the conference, provided a valuable compilation of analytical techniques. FAO has been the lead agency in a number of working groups in GESAMP, e.g., Working Group on the Scientific Basis for Disposal of Wastes into the Sea (GESAMP, 1975). With the financial support of SIDA (Swedish International Development Agency), FAO has held several workshops on study and control of pollution in the marine environment in relation to protection of living resources. Lectures presented at these workshops have been published and provide useful documentation of methods and measurement of pollutant concentrations and effects or techniques of observation on the state of marine pollution in a particular region (e.g. FAO, 1978).

International Council for the Exploration of the Sea

Having had a long tradition of international regional studies on the marine environment and its living resources, ICES continues to be a leader integrating hydrography and fisheries studies in the Northeast Atlantic. Canada and the United States are now members, and theoretically all the North Atlantic can be covered in ICES activities. ICES has an environmental quality committee with a number of working groups on marine pollution problems. Background documents on pollution of the Baltic and North Seas, as well as reports on monitoring of the North Sea, have been prepared. The ICES Working Group on Pollution Baseline and Monitoring Studies in the Oslo Commission and ICNAF areas recently prepared two reports on the results of studies on the selection and monitoring of dumpsites (ICES, 1978a,b). A subgroup of this working group reviewed from literature information the feasibility of monitoring biological effects of pollutants (ICES 1978c). This study led to the decision to hold an ICES Workshop on the Problems of Monitoring Biological Effects of Pollution in the Sea in Beaufort, North Carolina, in March 1979 (McIntyre and Pearce 1980).

CURRENT AND FUTURE CONSTRAINTS ON INTERNATIONAL ACTIVITIES TO CURB MARINE POLLUTION

Changing Priorities

The environmental ethic is rapidly losing ground to new issues such as energy, food supplies, inflation and unemployment. Environment becomes rather secondary when food supplies and energy are scarce. Although labour is concerned about environment, the risk of unemployment because a plant is shut down owing to the pollution it causes is sometimes a deterrent to environmental action during a period of high unemployment. In developing countries there continues to be some suspicion that wealthy industrialized nations are endeavoring to slow down development in underdeveloped countries under the guise of environmental protection. Developing countries sometimes feel that pollution is only a problem in the industrialized nations.

The slowdown in the popular environmental movement and the emphasis on energy has caused national treasuries to re-examine priorities. Environmental quality can seldom be "sold" on the basis of a need, in the way food, fuel oil, clothing or construction material can be sold. Emphasis in recent time has turned to resource matters. National allocations are shifting from environmental needs to resource production needs. International agencies are beginning to sense the national retrenchments in environmental areas. Peter Thatcher of the UN Environmental Programme recently expressed his concern that the MEDPOL program is being threatened with a serious curtailment, if not a total cessation of some components, because of lack of funds from participating states.

There has been hope that the UNEP Regional Seas Programme would have a high profile, because it covers parts of the ocean bordered by developing countries desperately yearning to receive the technology for marine studies from the developed countries. However, unless the Regional Seas Programme promises to increase food production from the sea, it is not likely to get much support from the developing countries. With withering resources in the UNEP pot, it is not likely that there will be much support from international funds to accelerate these regional programmes. National agencies providing funds for international development, such as SIDA in Sweden, CIDA in Canada and USAID in the United States, have reached their limit in annual allocations and are not likely to be of much help, with possibly some specific exceptions, in the Regional Seas Programme.

Changing Attitudes on Protection of the Ocean against Pollution

There have always been certain segments of society which have stated that the ocean should be used for waste disposal. Their contention is that the ocean has a certain assimilative capacity, and like a septic tank and sewage tile field serving a farm house, can accept a certain amount of waste without damage. Unfortunately, although this may be true to some extent, the assimilative capacity of some of our inshore waters has been exceeded, and rather unpleasant conditions with fish kills and other undesirable effects have occurred in some cases. Suddenly we became aware of the fact that the capacity of the oceans to accept waste is not limitless. We were less aware of the fact that natural transporting and mixing processes in many coastal waters are inadequate to safely disperse the effluents we release. We were even less cognizant of certain cumulative processes that occur in coastal waters, physically, chemically and biologically. Sometimes, tragic surprises occurred, such as Minamata disease in Japan as a result of consumption of mercury-rich shellfish.

During the late 1960s and early 1970s, it was a popular belief that the oceans are our last bastion of defense and should be protected at all cost. Partly for this reason, national agencies have resisted authorizing additional disposal of wastes into the sea under new legislation, and have moved to phase out existing disposal operations.

More recently, as land disposal options are being ruled out because of economic or technical infeasibility, the basis for banning the disposal of certain types of wastes into the sea has been increasingly questioned. In countries like the United Kingdom and other European nations, where land is in short supply, the use of coastal seas for waste disposal was never greatly curtailed, nor seriously considered for curtailment, even at the peak of national and international environmental concerns. There now is a movement in some countries, the United States included, to exempt certain waste disposal operations, e.g. urban municipal wastes in some coastal communities, from regulations such as secondary treatment, established under national and state legislation in the early 1970s. Deadlines for cessation of ocean dumping of wastes are being extended. This movement will undoubtedly accelerate and expand to other types of wastes in the 1980s. There will be more converts added to the roster of proponents, who feel that the ocean is a legitimate environment into which to dump wastes. This will unquestionably have an impact on the international environmental outlook, and adversely affect the availability of funds for marine environmental projects.

Scientific Uncertainty in Cause-Effect Relationship in Marine Pollution

Ecosystem relationships in the marine environment are complex and difficult to fully understand. Definitive studies on the biological effects of marine pollution, aside from standard bioassays, are time-consuming, costly and require a high degree of scientific competence and sophisticated experimental capacity in the laboratory and field. The financial support for such rather fundamental research is difficult to obtain. More often available financial and scientific resources are dissipated on short-term projects having an immediate application.

The result of this kind of allocation of resources to marine pollution problems is that authoritative information is unavailable on important cause-effect relationships. Speculation, rather than quantitative experimental laboratory results, is used to support a particular case for pollution control. Such evidence can often be readily disputed in a scientific forum, a public hearing or court of law. It is obvious that without convincing cause-effect data it is usually difficult, if not impossible, to achieve successful waste management and to effectively prevent environmental degradation.

To effectively support existing legislation on marine pollution control, it is essential to bring to bear on the identified problems some of the most imaginative and creative scientists available. The best analytical technology must be applied in both the laboratory and field to establish cause-effect relationships. The research has to be of a long-term nature to provide a full picture of the ecological impact of a substance over a long period of time. "No-effect" levels must be experimentally derived.

Unfortunately, many competent scientists are shifting from research on the applied marine pollution problems to more fundamental pursuits, as additional funding for basic marine research becomes available. Some of this scientific information may have application to marine pollution problems, but not necessarily so. In the meantime, many pressing issues on marine pollution go unresolved for lack of good scientific information. It becomes difficult for decision-makers to ascertain whether a particular marine waste disposal operation is creating a problem or not. Anti-pollution legislation and regulations become difficult to administer and enforce. Pollution may continue and eventually lead to serious environmental degradation, because there was no sound scientific basis on which to control it in the first place. Existing international conventions may not be properly administered and there may be inadequate scientific basis on which to prepare acceptable amendments or to draft new conventions.

The Popular Environmental Movement

The number and effectiveness of environmental groups are declining. Those individuals who merely joined the environmental movement because it offered a "cause" to be identified with in the late 1960s and early 1970s are finding other more current causes to support. The "hard-core" environmentalists are sometimes misled by misinformed zealots who can do a great deal of harm to the credibility of the environmental movement. Many environmentalists and conservationists are motivated by issues emotionally generated. They tend to ignore sound scientific data available on these issues. Their tactics in fighting a particular issue may be questioned, such as violation of laws on the fishing grounds or wilfully damaging private property.

The genuinely sincere environmental groups, representing concerned citizens on particular environmental issues, often do not have much popular appeal or financial support. They are rather ineffective as lobbying bodies in legislative assemblies. Because of lack of funds, any momentum that is generated to control a particular issue is often short-lived, and can be overcome merely by perseverance of a proponent. Lobbying by a body with sound financial backing may be less obvious through the media, but is usually much more effective than the lobbies of environmentalists in achieving a specific objective.

Internationally, the environmental and conservation movement may be represented by groups having world-wide recognition. The Friends of the Earth sometimes have observers at various international meetings on environmental matters, such as ocean dumping. They may make carefully prepared interventions on certain issues. But offsetting their influence are organizations representing the operational entities, with strong financial support, such as Oil Companies International Forum (OCIMF), International Association of Independent Tanker Owners (INTERTANKO), International Chamber of Shipping (ICS) and Oil Industry International Exploration and Production Forum. The rapid rise in marine environmental concerns during the late 1960s and early 1970s has been followed by a decline in the late 1970s (fig. 2).

CONCLUSIONS

Parallel with national environmental activities there has been a great deal accomplished internationally on understanding and controlling marine pollution globally. The conclusions that one can draw concerning these activities and the prospects for the future are as follows:

1. Oil pollution from ships is the main contaminant controlled internationally by conventions. IMCO administers these conventions but enforcement is carried out by member states.

2. There has been no success so far in bringing into force the International Convention on the Prevention of Pollution by Ships, 1973. A protocol developed in 1978, to cover the oil pollution aspects of this convention, is being promoted for entry into force soon.

3. The Convention on the Prevention of Marine Pollution by Wastes and Other Matter, 1972 (London Dumping Convention), has been in force since 1975, and national legislation of ratifying states has been generally brought into line with this convention. The regional Convention for the Prevention of Marine Pollution by Dumping from Ships and Aircraft (Oslo Dumping Convention) generally takes precedence over the London Dumping Convention for those states ratifying both conventions, and it is difficult to amend the latter before the former is appropriately amended.

4. Land-based sources of marine pollution, including effluents released through ocean outfalls, river-borne discharges and atmospheric

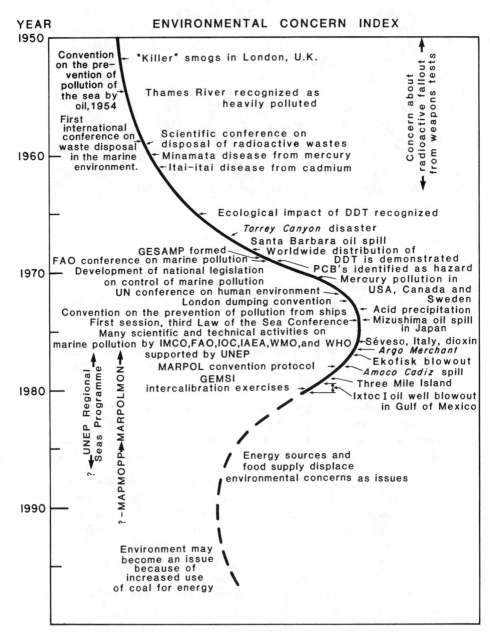

Fig. 3.2 Changing level of public concern about the marine environment, as based on significant environmental events.

emissions, are not likely to be controlled soon through international conventions. Extension of national jurisdiction recently by many nations to 200 nautical miles has further widened the coastal zone in which pollutant discharges, e.g. those from drill rigs in exploitation of oil, will not likely be controlled globally by international agreement.

5. Pollution from exploitation of the sea-bed in truly international waters, e.g. manganese nodules in the eastern equatorial Pacific, representing resources that are "the common heritage of mankind", will be probably controlled by the International Sea Bed Authority, if and when the new Law of the Sea comes into effect.

6. International agencies have endeavoured through conferences, workshops, seminars, panels and working groups to fulfil a role in understanding the problems of marine pollution, pertaining particularly to their mandates and responsibilities in the marine environment: (a) IMCO — pollution prevention from ships; (b) FAO — protection of living resources of the sea against pollution; (c) Unesco — marine sciences and understanding processes related to marine pollution; (d) WMO — atmospheric pollution and air-sea exchange of pollutants; (e) WHO — protection of human health from pollution of marine beaches and seafood; (f) IAEA — impact of radioactivity on the marine environment and living resources; (g) UN — effects of sea-bed exploitation and coastal development on the marine environment; and (h) UNEP — all environmental aspects of marine pollution. Synthesis of the knowledge and needs for information of the various UN Agencies on marine pollution come together in an interdisciplinary way in GESAMP.

7. The Regional Seas Programme of UNEP, covering 10 regions, has promise for developing a number of comprehensive regional marine studies: (a) Mediterranean; (b) Red Sea and Gulf of Aden; (c) Kuwait Action Plan Region; (d) Caribbean; (e) West Africa; (f) East Asian Seas; (g) South-East Pacific, and (h) South-west Pacific; (i) East African Region; and (j) South-west Atlantic. Action plans are in effect for (a)–(c), and in various stages of development for the others. Emphasis has been placed on transferring technology in marine pollution measurement and assessment of effects to developing countries.

8. The Intergovernmental Oceanographic Commission's IGOSS Pilot Project on Marine Pollution (Petroleum) Monitoring (MAPMOPP) 1975–1980, covering the North Atlantic Ocean by ships of opportunity, has been considered a success in observing oil slicks and

measuring the amount of oil residue as tar balls. The interpretation of measurements of dissolved/dispersed hydrocarbons in the surface layer of seawater by UV-Spectrofluorescence has been questioned. MAPMOPP is being phased into a more permanent activity in mid–1980, Marine Pollution Monitoring (MARPOLMON), with oil pollution monitoring continuing to be the main emphasis. It is doubtful that the system will ever be used for monitoring other pollutants in the sea.

9. The national decline in emphasis on environmental concerns is reflected internationally with a slowdown in new initiatives on marine pollution investigations and some curtailment in ongoing programs. With energy and food-related initiatives displacing environmental projects, the 1980s could witness increased use of the oceans for waste disposal and a return of international agencies to resource and energy orientation.

References

FAO, 1971a. Report of the seminar on methods of detection, measurement and monitoring of pollutants in the marine environment. *FAO Fish. Rep.* Vol. 99, Suppl. 1.

FAO, 1971b. Report of the ACMRR/SCOR/ACOMR/GESAMP Joint Working Party on Global Investigation of Pollution in the Marine Environment (GIPME), San Marco di Castellabate and Rome, Italy, October 11–18, 1971. *FAO Fish. Rep.* Vol. 112.

FAO, 1978. Lectures presented at the Fifth FAO/SIDA Workshop on Aquatic Pollution in Relation to protection of Living Resources. Scientific and Administrative Bases for Management Measures, Manila, Philippines, January 17–February 27, 1977. Food and Agriculture Organization of the United Nations, Rome, Rep. TF-RAS 34(SWE) – Suppl. 1.

GESAMP, 1975. Scientific criteria for the selection of sites for dumping of wastes into the sea. IMCO/FAO/UNESCO/WMO/WHO/IAEA/UN Joint Group of Experts on the Scientific Aspects of Marine Pollution – GESAMP. Reports and Studies No. 3.

Goldberg, E. D. 1976. *The health of the oceans,* The Unesco Press: Paris. 172 p.

ICES. 1978a. Report of the International Council for the Exploration of the Sea to the Oslo Commission, the Interim Helsinki Commission, and the Interim Paris Commission, 1977. International Council for the Exploration of the Sea, Charlottenlund Slot, DK-2920 Charlottenlund, Denmark. *Coop. Res. Rep.* No. 76.

ICES, 1978b. Input of pollutants to the Oslo commission area. *Ibid.* No. 77. 57 p.

ICES, 1978c. On the feasibility of effects monitoring. *Ibid.* No. 75. 42 p.

IOC, 1976. *A comprehensive plan for the global investigation of pollution in the marine environment and baseline study guidelines.* Unesco, Intergovernmental Oceanographic Commission, Tech. Ser. No. 14.

McIntyre, A. D., and J. B. Pearce, eds, 1980. Biological effects of marine pollution and the problems of monitoring. Rapp. p.-v Réun. Cons. int. Explor. Mer. 179: 1–346.

Ruivo, M. 1972. *Marine pollution and sea life.* London: *Fishing News (Books) Ltd.*

Waldichuk, M. 1973. International approach to the marine pollution problem. *Ocean Management, 1:* 211–261.

Waldichuk, M. 1978. *Global marine pollution: an overview.* Unesco, Intergovernmental Oceanographic Commission, Tech. Ser. No. 18.

WHO, 1975. Ecological aspects of water pollution in specific geographical areas: study of sublethal effects on marine organisms in the Firth of Clyde, the Oslo Fjord and the Wadden Sea. Long-Term Program in Environmental Pollution Control in Europe. Report on a working group convened by the Regional Office for Europe of the World Health Organization, Wageningen, Netherlands, December 2–4, 1974. World Health Organization, Regional Office for Europe: Copenhagen, Denmark.

WHO, 1978. *Health implications of nuclear power production.* Report on a Working Group, Brussels, December 1–5, 1975. World Health Organization, Regional Office for Europe, Copenhagen, Denmark, EURO Reports and Studies No. 3.

WHO, 1979. *Environmental health impact assessment.* Report on a WHO Seminar, Argostoli, Kefalonia, Greece, October 2–6, 1978. World Health Organization, Regional Office for Europe, Copenhagen, Denmark, EURO Reports and Studies No. 7.

Discussion

Comment: I would like to disagree completely with Dr. Waldichuk on one issue that he raised. I think we are all cognizant of the fact that scientists do not relate very well to the public, but I also think the politicians do not relate well either. I feel that when we have asked the public systematically, not asked the politicians, the public is still as interested in paying for environmental management as it always was. I think the politicians are the people that are backing away.

Waldichuk: I think that you are quite right, but, in a democratic society the politicians are supposed to respond to the public interests. I do not think the public has insisted on the politicians responding in the way that they should, if, in fact, the public feels that way.

Comment: It is worth remembering the GESAMP definition of pollution. They call it "the damaging excess," meaning if there's no damage, there's no pollution. They very specifically said that "change does not constitute pollution."

Comment: I would like to comment upon the statement that the Oslo Convention prevented any kind of progress in the London Dumping Convention. The truth is that the Oslo Convention is as diverse as the London Dumping Convention. I know that some countries in the London Convention used the excuse that an amendment in the Oslo Convention has to go forward for the Oslo Convention. But, I think it needs to be stated that there are progressive countries in Europe too.

Regarding the comment that incineration should be looked upon a little bit closer, maybe we should have looked a little bit closer and added appropriate provisions to the conventions two or three years ago. In a forum like this, it has to be pointed out that the United States was one of the key countries lobbying for that amendment.

Waldichuk: I think that, in general, we have to realize that regional conventions will have precedence over a global convention if a country belongs to two conventions. This is natural in as much as it is generally felt that the regional convention should be the more stringent one.

The other point I should make is that we have talked this morning about getting combustion products into the atmosphere from incinerator ships without fully knowing the consequences. I should state that the Scandinavian countries, in general, including Denmark, are totally opposed to the continuing incineration of substances in incinerator ships. To some extent, I sympathize with them and will continue to do so until we know more precisely what effects the combustion products are having on the atmosphere and the sea downwind. We really do not know what the emissions are and how they behave when different substances are incinerated on incinerator ships under a variety of environmental conditions.

CHAPTER 4

Commentary
Ocean Science: The Linchpin of Ocean Management

JAMES W. CURLIN
Department of the Interior
United States

I have resisted the temptation to delve into the historical evolution of science policy and ocean policy from the post-war era to the present, although to fully comprehend where we presently are requires that this be done. Suffice it to say that we have been unsuccessful as a nation in providing an integrating theme for focusing U.S. ocean policy, and consequently for providing a framework for marine science in a policy sense.

The post-war era of the late 1950s and 1960s, with its emphasis on technology and space exploration, focused ocean programs on marine research and development, with a strong bias toward military application. During the 1970s, with the public's faith in technology waning and its preoccupation with environmental protection at a zenith, federal ocean activities shifted toward regulation and control of ocean use. We have also flirted with the problems of international ocean space in the United Nations Conference on the Law of the Sea. In the meantime, marine science has followed the whims of the federal R&D budget with no clear indication of the expected endpoint.

I should make it clear at this point that my remarks apply solely to the directed, application-oriented research funded by the federal government and not to fundamental research, funded primarily through the National Science Foundation.

The *Federal Plan for Ocean Pollution Research, Development, and Monitoring,* which was compiled by the Committee on Ocean Pollution Research Development, and Monitoring (COPRDM) and released in December 1979, provides graphic evidence of the shortcomings in dealing with marine science in general, and pollution research in particular. I suggest that there are three factors that must be considered in developing a more rational approach to marine science in the United States:

1. An overarching principle for focusing marine science and technology in the context of U.S. ocean policy;
2. Fundamental changes in the way we conceive, structure and execute ocean studies;
3. Significant changes in the way marine science is organized, administered, and coordinated at the federal level.

Each of these factors relates to one another, and collectively they can determine the atmosphere, and therefore the effectiveness, of federally funded application-oriented marine science in the future.

FOCUSING MARINE SCIENCE

In the past, we have tried to peg ocean policy on science and technology, environmental protection, and international law. All of these have failed to provide an integrating theme upon which to build a lasting framework for ocean policy. U.S. ocean policy is a grab bag of single-purpose laws, each of which fails to acknowledge the coexistence of other similarly well-intentioned laws and other competing uses in the ocean.

In the near future we will extend our national control over 1.4 billion acres of ocean space and marine resources within a 200-mile exclusive economic zone. This zone and the resources therein will for the first time become legal property of the United States, free of any foreign claims save the right of peaceful ingress and egress. This will happen in one of three ways: (1) through successful negotiation of a ratifiable treaty in the LOS Conference, which has already reached agreement on the economic zone; (2) by extension of national jurisdiction through legislative action; or (3) by Presidential proclamation, as was done in 1945 by President Truman. With the advent of a U.S. economic zone, a foundation will be laid for managing the waters, seabed and the resources of the economic zone as integral units in the same general manner that we treat the public lands under the stewardship of the federal government. Instead of exercising control over the coastal waters in a piecemeal fashion, as has been the policy of the past, the United States will have an opportunity to develop a management system based on the precepts of multiple use and sustained yield.

For the first time there will be a focus for U.S. ocean policy aimed at comprehensive resource management with an emphasis on development of an ocean and coastal management system, which can ensure the balanced use and protection of the resources within the economic zone.

Science and technology are the working tools of resource management. In order to fashion a workable ocean and coastal management system, it will be necessary to learn a great deal more about the structure, functions, transfers, transports, synergisms, antagonisms, processes, profiles, interactions, and responses of the various ocean ecosystems. Scientific knowledge must be the basis for making rational decisions on alternative uses and for resolution of conflicts among the users of the ocean system. The knowledge provided by the scientific community will have a direct application in the systematic management of the economic zone.

While I do not wish to diminish the importance of "blue water" oceanography, it is increasingly apparent that oceanographic efforts within the 200-mile economic zone and in the outer continental shelf will tend to attract an even greater proportion of the federal ocean research budget in the future. Circulation over the continental shelf will likely displace studies of coastal upwelling in far flung regions of the world. It will be easier to fund geological investigations of the continental slope than to study spreading centers in the mid-ocean ridges. Distant water oceanographic research is becoming increasingly more expensive because of escalating energy costs and the general inflationary trends, while the results of such studies — sound in principle and scientifically credible though they are — fall short of the directly applicable resource information that titillates government budgeteers. While this attitude may be considered another manifestation of bureaucratic shortsightedness by some, it nevertheless reflects the realities of the prevailing attitude. Clearly, we must ensure that global oceanography is kept at an acceptable level; but pragmatically, if an additional research buck is to be turned for the ocean, it must be done on the justification of adding knowledge for the management of our coastal waters.

STRUCTURING OCEAN STUDIES

The ocean is a textbook example of a natural ecological system. As such, it must be studied in a systematic fashion. Within the 200-mile economic zone the linkages among components of the system, particularly at the boundary layers, become controlling. To fully understand the processes and interactions within the ecosystems of the coastal ocean requires holistic interdisciplinary studies which join the scientific talents of several disciplines into a research team. Unfortunately, oceanographers, for one reason or another, have had limited experience in interdisciplinary ecosystem research.

We must not only learn to measure and model interrelated aspects of chemistry, biology, ocean physics, and geology and geophysics, but we must also generate the atmosphere and opportunities for interdisciplinary skills to develop and exist. Nor can we ignore the importance of socio-economic research to a well-found ocean and coastal management system. However, we face both institutional problems and conflicting attitudes among the disciplines in bringing this about.

Interdisciplinary ecosystem research requires careful planning, effective coordination and competent management. Few research administrators presently have the management experience and skills necessary to carry off large-scale studies successfully. The COPRDM Federal Plan and its supporting documents confirm that few holistic ecosystem studies related to ocean pollution have been undertaken.

Scaling is a factor in ecosystem research, both in temporal and spatial terms. Sophisticated planning and coordination is required to ensure that information collected by one discipline is available in a suitable form as input to another discipline in an appropriate timeframe. This kind of PERT charting and scheduling, while familiar to an engineer or production manager, is foreign to most science administrators.

Another limitation on interdisciplinary ecosystem research is imposed by institutional funding. Many regional pollution studies will require multi-agency funding and participation to ensure that a critical mass of scientists and funds are available for meso-scale studies. Yet funding and the allocation of personnel for government-wide pollution research and monitoring is left to the arcane vagaries of the authorizing-budgeting-appropriations processes. Clearly, if interdisciplinary ocean ecosystem research is to be carried out successfully, adjustments will have to be made in the way that ocean research is conceived, planned, funded and organized in the ethereal world of government policy makers.

SHORTCOMINGS IN THE PRESENT APPROACH

The *Federal Plan* identified approximately 1,000 individual projects dealing with pollution research, development and monitoring, which are conducted by seven departments and four agencies. While dealing specifically with pollution-related activities, the COPRDM assessment is representative of ocean science as a whole. Marine research programs are structured in response to the missions of the individual agencies. If the composite research program of these agencies reflects the proper mix and complement needed in oceanographic studies, it is because of happenstance and not by design.

With the exception of the recent activities of COPRDM mandated by the Ocean Pollution Research and Development and Monitoring Planning Act of 1978, there are no mechanisms for cross-cutting the agencies' budgets to get a clear picture of the fiscal resources being committed to marine R&D. More importantly, there are no effective processes within the federal government to ensure that the ocean science which is conducted by one agency meshes with the ocean studies of its sister agencies.

Projects are developed within each agency, budgets are formulated in isolation, and the Office of Management and Budget (OMB) deals with them in a like manner. Within the Congress a similar scenario is repeated. Appropriations are dealt with line item by line item without evaluating the range and content of the related ocean programs of other agencies.

While the National Ocean Pollution Research and Development and Monitoring Act of 1978 provides for the preparation of a five-year plan for budget planning and coordination of research, it nonetheless provides no assurance that such a plan will lead to better coordination of overall research among the agencies or that studies will be organized to provide the integrated knowledge about ocean ecosystems that is needed to arrive at measured decisions.

It is clear that more will be required than a mere "plan" if we are to provide the guidance and coordination that are required to meet the needs for managing the U.S. Economic Zone. "Lead Agency" concepts, interagency plans, and good faith cooperation among the agencies will accomplish a modicum of coordination, but we must not fool ourselves into thinking that these approaches will provide the structure and the governmental devices necessary to ensure a balanced marine science program for the future.

Our past performance in dealing with the functional problems of providing the information needed for an accelerated OCS oil and gas leasing program which was foreseen as early as 1973, and the still scanty management information we have for managing marine fisheries, and more directly the insufficiencies of useful knowledge to deal with the subject of this book — marine pollution — is *prima facie* evidence that we are not conducting marine science in the manner necessary to meet the needs of resource managers.

Bold strokes will be needed in the future, rather than the subtle body language we have been using up to present, if we are to make the necessary adjustments within the federal establishment to meet the scientific needs for managing the future economic zone of the United States. These adjustments include: (1) policy framework for formulating national goals for federal ocean R&D; (2) a well developed strategy for achieving these national goals; (3) a stable base of funding for ocean R&D; (4) changes in our institutional approaches to marine science in order to deal with the ocean in a systematic way; and (5) mechanisms within the federal establishment to ensure a collaborative approach to ocean science among government agencies, academics and the private sector.

Discussion

Comment: I think there have been very difficult economic trade-offs made in terms of quality and pollution problems and that there has been some decline in the interest in pollution as a major issue. After the London fog and the Minamata Disease, there was not all this major effort. I think we have learned how to live with the fog, with the fact that shellfish beds are going to be closed, and that you cannot swim in the East River. These are the kinds of things that one just cannot afford to do because of the trade-offs.

My concern, however, is that we make some suggestions about how to get more involved and how to get more money into the systems. It may work. But it seems to me that in the long term we have a different kind of problem in pollution. It is not what we know about it; it is what we do not know about it. Sooner or later we may wake up and find chlorine in the atmosphere or some other problem. We now have a situation in the ozone layer that we will have to live with for either decades or centuries. Perhaps a more deadly issue is lead poisoning. We must face the unknown aspects of pollution. When we look at the major pollution problems of the future, we must not only examine what we really know and what we can accept, but also evaluate the things that we do not know and then try to predict future problems.

Comment: I wanted to follow a point that both Dr. Waldichuk and Dr. Curlin made. Regarding the graph which shows the declining interest, I think that this is a mistaken and an unfortunate way to approach our future and the need for scientific research. I think that Dr. Curlin displayed a misplaced emphasis by indicating that we need to refocus our thinking on the territorial seas in terms of development, rather than in terms of environmental protection. The people in the United States have not in any way said that they are less interested in clean air, clean water, or a safe environment. I think that they continue to have that concern, but they are more sophisticated now. The early 1970s saw the passage of environmental laws, but the much tougher issues of the regulations and the management are now before us as part of the regulatory process. I think it is mistaken to say that there is a downward interest. Both the Environmental Protection Agency and the Council on Environmental Quality have studies which show continuing interest. To try and shift focus away from that does harm to the kind of concern that I think is captured in the theme of this conference; that is "What are the impacts on society?"

In regard to the international area, which Dr. Waldichuk discussed, I see some real problems because increasingly we are looking at problems of a global nature. The United States will soon make public the Global 2000 Study, which looks into problems not only of ocean pollution but also of the whole range of environmental concerns. The example that Dr. Waldichuk talked about briefly, the London Dumping Convention, troubles me. It seems to me that there has been a great deal of difficulty in accomplishing any real monitoring and assessment of the impacts of substances such as radioactive waste. The Northeast Atlantic dump site has never really been monitored. There are continuing problems about setting up an effective system of monitoring. This is partly because only three or four countries do it and the other countries are not that interested in participating in such studies partly because it is difficult to reach a consensus on working and arriving at solutions. It seems to me, at an international level, there needs to be much more of an effort above and beyond the scientific aspect. We must try and address the political problems of how to deal with issues such as ocean dumping of radioactive wastes, and answer questions such as: Do you set up a technical coordinating committee, like that which is in place for tanker safety issues? Could you comment on how, in a technical sense, one can establish a mechanism for educating those concerned and reaching for some solutions for an issue such as radioactive waste.

Waldichuk: I would like to comment first about declining interest. I think the problems are no less than they were in the sixties or early seventies. We just have to face the fact that the media, perhaps, are not relating to the public the kinds of problems that exist today. If we do not recognize the fact that we are in a period of decline, we are merely burying our head in the sand.

Concerning the matter of monitoring of, for example, the Northeast Atlantic dump site for radioactive wastes, I think that, in general, we are up against some of the traditional resistance that has existed in Europe and particularly in the United Kingdom against long-term monitoring. We have to demonstrate to them that by monitoring you can actually learn something useful and then try to gain cooperation from all the different users of the dump site. It is partly an educational process.

UNIT TWO

Accidental Pollution: A Case Study of the *Amoco Cadiz* Oil Spill

Photo courtesy of Eric Schneider

IN MARCH OF 1978 THE SUPERTANKER *AMOCO CADIZ* RAN AGROUND OFF THE northwest coast of Brittany, France. Efforts to free the tanker were unsuccessful, and because of the extensive press coverage, the entire world witnessed what was the largest oil spill in maritime history. During the 18 days following the grounding, the entire cargo of some 223,000 tons of Middle Eastern light crude oil and the bunker fuel were spilled into the ocean.

The purpose of this unit is to provide a case study of the *Amoco Cadiz* oil spill. Before we look at the specifics of the spill, however, it is useful to put the *Amoco Cadiz* case and accidental pollution from oil tankers in general in perspective.

The table on facing page lists tanker originated oil spills greater than 25,000 tons over the period 1967 to 1979. As you can see, there were 24 oil spills over 25,000 tons during the period, and the *Amoco Cadiz* spill was by far the largest. Thus, large oil spills from tankers are by no means isolated events and, in fact, there was at least one spill greater than 25,000 tons every year from 1967 through 1979. A similar table, including all spills greater than 5,000 tons, would show at least 50 reported in the period through 1978. So large oil spills are not rare, and at least one occurs every year.

The world map shows the geographic distribution of tanker oil spills, indicated by the black dots. It is taken from a recent IMCO report, and indicates the location of spills greater than 5,000 tons through 1978. It is interesting to look at the geographic distribution of large spills. We see that a great many of the oil spills are concentrated along the coast of industrialized countries or along the traffic lanes leading to industrialized countries. Thus there is a pattern; marine pollution is a price paid for high levels of economic activity.

Environmental disasters like the *Amoco Cadiz* raise important scientific, social and economic, and national and international policy issues. From a scientific point of view, there is an important need to understand better the short-run effects and the longer-term impacts of oil spills on the natural environment. As we were reminded this morning, however, it is important not only to try to understand the impacts of oil spills and marine pollution from a scientific perspective, but to convey these findings to policy makers and to the public in a way that can be understood. Dr. Laubier will discuss the ecological impacts of the *Amoco Cadiz* oil spill in Chapter 5.

From a social and economic perspective, it is important, for a variety of reasons, to understand the costs involved with oil spills. For one thing, there's the issue of compensation. In the period following environmental disasters, such as the *Amoco Cadiz,* public debate takes place in an emotion-charged arena, with claims and counter claims being made regarding the extent of damages. Compensation clearly depends on an assessment of damages, which requires economic studies to determine costs.

Having said the word "costs," we must define what we mean by economic costs. The actual cost of an oil spill may be somewhat different from that reported or made available in public debate. We also must ask the question: costs to whom? Oil spills like the *Amoco Cadiz* can result in damages to local tourism, commercial fishing and other regional activities. There are also damages to a broader area, such as clean-up costs borne by the nation as a whole. There are still other categories of damages, those that are borne by residents

MAJOR TANKER OIL SPILLS

	SHIP		TONS
1967	*Torrey Canyon*		100,000
1968	*Ocean Eagle*		45,000
	World Glory		45,000
1969	*Keo*	c.	25,000
	Pacocean		30,000
1970	*Ennerdale*		40,000
	Chryssi		31,000
1971	*Wafra*		64,000
	Texaco Oklahoma		31,500
1972	*Sea Star*		65,000
1973	*Napier*		30,000
1974	*Metula*		56,000
1975	*British Ambassador*		50,000
	Spartan Lady		25,000
1976	*St. Peter*	c.	30,000
	Cretan Star		28,500
	Argo Merchant		32,000
1977	*Irene's Challenge*		36,000
	Caribbean Sea		35,000
1978	*Amoco Cadiz*		230,000
1979	*Betelgeuse*		25,000
	Aegean Captain/Atlantic Empress		145,000
	Burmah Agate		31,800
	Independente		86,000

SOURCES: 1967–1978, IMCO News, No. 1 of 1979
        1979, Center for Short-Lived Phenomena

of the world, such as the value of the lost oil, the value of the tanker, and the research and legal costs that are involved in dealing with oil spills. The point is we want to be very careful in the way we define the costs of oil spills and we are concerned about the distribution of costs across groups. A careful examination of the costs of marine pollution will allow us to understand better the impacts of pollution and therefore the potential benefits — from reducing pollution. The concept of economic costs of oil spills, applied to the case of the *Amoco Cadiz,* will be explained more fully in Chapter 6.

From a public policy point of view, there are a number of important issues that arise which underscore the importance of understanding the effects of oil spills. There are policies that are under consideration to improve the safety of tankers or to change shipping lanes. In the United States, we frequently have delays or safeguards imposed on offshore drilling to mitigate, reduce or prevent oil spills. Often there's a need to plan for the location and investment in facilities to deal with oil spills should they occur. In all of these cases, there is the perception that costs will arise from oil spills, and often there is the presumption that the benefits to be realized (damages avoided) justify the costs incurred to avoid the threat of damages. However, there have been few comprehensive damage assessment studies. Thus, there's a need to understand the costs of

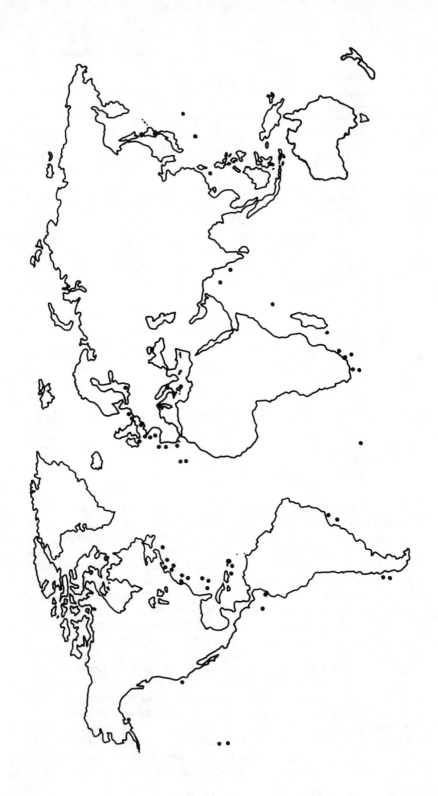

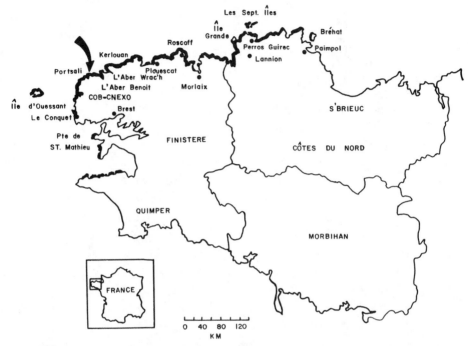

marine pollution and how significant they're likely to be. This means, then, that we need to improve our scientific base of information and our economic methodologies for assessing damages. Another public policy issue has to do with international conventions dealing with oil spills. These will be discussed in Chapter 7.

It is useful to provide background information on the *Amoco Cadiz* oil spill based on a NOAA/EPA report of the spill. About 32% of the oil (74,000 tons) is believed to have evaporated, 33% (76,000 tons) was lost at sea in one form or another, and some 35% (80,000) came ashore. The second map shows the Brittany region that was affected by the *Amoco Cadiz* oil spill. Brittany is the western-most part of France, extending out into the Atlantic. In the lower left hand corner, you can see Brittany in relation to the rest of France. The geographic extent of the spill is indicated by the darkened area, extending roughly from Paimpol and Brehat around the northwest coast. Over 300 kilometers of coastline were affected by the spill. The arrow indicates the area off Portsall where the *Amoco Cadiz* actually ran aground.

Thomas A. Grigalunas
University of Rhode Island
United States

CHAPTER 5

Ecological Impacts

LUCIEN LAUBIER
National Center for the Exploitation of the Oceans
France

INTRODUCTION

During the night of March 16 and 17, 1978, the supertanker *Amoco Cadiz* ran aground 1.5 nautical miles from the shore at Portsall, on the northwest Brittany coast. Almost the entire cargo, with a volatile fraction estimated at 30% to 40% of the total, was lost to the sea within a rather short period of about 14 days. Approximately 300 kilometers of different types of shores have been more or less polluted.

The general area affected by oil pollution is characterized by an alternation of sandy beaches and rocky areas of several types of granite. Due to the high tidal ranges (from five to nine meters), very strong tidal currents occur generally parallel to the coastline. Special marine biotopes also exist in the area: the "abers" which are submerged estuaries of small rivers about ten to fifteen kilometers long and no more than one kilometer wide, with sandy and muddy areas; several small bays with muddy bottoms, especially in the Bays of Morlaix around Carantec; and several marshes, the largest one being the Lannion marsh in the eastern area.

Generally speaking, the marine ecosystems of the area are very well known. Due to the research work carried on since 1871 by the biological station at Roscoff, the intertidal and subtidal benthic ecosystems are precisely defined qualitatively and quantitatively from the Santec Beach up to the eastern part of the Bay of Lannion. West of Santec, recent quantitative data on high energy sandy beaches are also available. The lesser known areas are the abers and the western part of the coast. From an ecological point of view, the area polluted by the *Amoco Cadiz* oil spill has a very high diversity within the different communities, some of them having high productivity (e.g., the *Laminaria* belt).

Oil spills in the same area have occurred three times within the past 11 years: in May, 1967, some 20,000 tons of weathered oil from the *Torrey Canyon* wreck came ashore on the coast between Lannion and Paimpol; in February, 1976, a large but empty tanker, the *Olympic Bravery* ran ashore on the north west coast of Ushant Island, spilling some 1,400 tons of bunker oil; in October of the same year, a small East German tanker, the *Boehlen,* ran aground on the north-west of the Ile de Sein carrying 9,600 tons of heavy Venezuelan Boscan crude, and it was estimated that about 5,000 tons spread into the marine environment within months. None of these oil spills were comparable to what occurred in the case of the *Amoco Cadiz,* but some studies were performed by French investigators, in which they gained a limited experience with the problem.

THE STUDY PLAN

Immediately after the wreck, on March 17, the Ministry of Environment required CNEXO to prepare a long-term program to assess the ecological impact of the hydrocarbons on the marine environment. Several French academic laboratories (Universities, National Museum of Natural History) and scientific organizations (National Geographic Institute, French Institute of Petroleum, Scientific and Technical Institute of Marine Fisheries, National Center for the Exploitation of the Oceans) are participating in this program. Besides the French effort, there has been a large foreign contribution including: from the United States, the National Oceanic and Atmospheric Administration and its university associates and the Environmental Protection Agency; and from Canada, the Bedford Institute of Oceanography. Special funding for the main French program was provided by the Ministry of Environment, and a contribution from the Amoco Company supported the development of joint French-American programs. A leading scientific committee for the main program and a joint scientific commission for the complementary French-American program were set up to follow the progress of the research work and develop the lines of study on a three-year basis.

The main program prepared for the three year period was divided in three parts:

1. Monitoring the chemical quality of the water, the sediments and the marine organisms. This included a short-term study mapping the oil at sea and on the shore at different periods using infrared and colour pictures taken by airplane, and long-term studies especially in the low-energy muddy sediments of the abers.

2. Studying the ecological effects of the spill, especially on exploited species such as the algae (kelp and red algae), crustaceans, fishes and oysters. This part included a short-term study to collect, identify and count the dead animals on the shores, and a long-term survey of the resulting effect and the conditions of the restoration of the most heavily affected communities.
3. Studying the microbiological processes associated with biodegradation. Due to the limited available number of marine microbiologists in France, this program was rather restricted, including long-term studies of biodegradation rates in some highly polluted environments such as the marine marshes.

The implementation program developed through the French-American joint scientific commission included three main parts:

1. Detailed quantitative evaluation of the restoration of the subtidal benthic ecosystems;
2. Study of microbiological processes and rate of biodegradation of the oil;
3. Detailed studies of special marine biotopes and possible control of their restoration, the abers and the marine marshes.

Twenty months after the wreck, an international symposium was organized in the Centre Oceanologique de Bretagne by the end of November, 1979, and some sixty papers were presented, most of them concerning the assessment of the ecological impact of the hydrocarbons on the marine environment. It is important to recall that even after the IXTOC I blow out, the *Amoco Cadiz* case remains the largest accident of oil polluting the marine environment.

GENERAL REMARKS

The *Amoco Cadiz* was carrying 223,000 tons of two types of crude oil, both of them "light petroleum," coming from Iran and Arabia. The aromatic fraction, considered the toxic fraction, represented a large percentage of the total oil content, about 30% for the Arabian type, and 35% for the Iranian type. It was a major accident, with regard to the quantity of oil spilled, and the rate at which the oil was spilled into the environment. This huge quantity of oil went into the sea in no more than 14 days: this rate is about ten times faster than the Ekofisk accident, which involved a similar type of oil, or the more recent IXTOC I blow out.

A third important point to recall is that the wreck was no more than one and a half miles from the shore and therefore the time of weathering of the oil at the surface of the sea before it went ashore was a few hours to a few days only. It seems that a large quantity of the aromatic fraction was still present in the oil when it touched the coast; and aromatic fractions are more toxic to the marine fauna than are other fractions.

The fourth point to underline concerns the meteorological factors which led to the dispersion of oil on the sea surface. The accident took place during the night of March 16 and 17, and usually, at that time, the winds blow from west to east and from north-west to south-east. Thus, during a two-week period, until the beginning of April, all the oil was transported eastward and touched the coast successively from west to east, up to a distance of about 150 kilometers from the wreck. The motion of the oil at the surface of the sea has been evaluated at about 2.5 to 3% of the speed of the wind in the same direction. This means that for a wind of 20–30 kilometers/hour, the current motion of the water has practically no effect on the displacement of the oil. The wind controls completely the displacement of the oil.

MAJOR SCIENTIFIC RESULTS

The chronology of the events of the spill can be divided into three phases: progression, stabilization and decontamination.

Starting from Portsall in the morning of March 17, the oil successively reached Aber-Wrac'h (March 19), Roscoff (March 20), the bay of Lannion (March 21), the bird sanctuary of the Seven Islands (March 22), and Sillon de Talbert (March 23). In the beginning of April, the reversal of wind direction pushed the oil to the west, and all places polluted during the first phase played the role of secondary sources of pollution. Le Conquet and Ushant Island were touched by the oil on April 11, the Raz de Sein on April 13, Douarnenez on April 22, and a small amount of oil went to the coast in the bay d'Audierne in May. By the end of April, this second phase of stabilization was completed, with the oil trapped in several areas. The third phase, decontamination, started around the beginning of May 1978 and is still going on more than two years later.

Pollution of the coast

At the maximum extension of the pollution (at the end of March), it has been estimated that 60,000 tons of oil polluted 72 kilometers of coast-line. In late April, 1978, 300 kilometers of coast-line were polluted although this figure decreased with time to 220 kilometers in November, 1978, 100 kilometers in March, 1979 after the winter storms, and nearly 50 kilometers in November, 1979. Using granulometry and level of energy of each area, a vulnerability index was established for the Brittany coast, with ten categories from rocky points polluted for a few weeks to marine marshes and muddy flats in the abers polluted for probably more than 10 years. Low energy and fine particle size are directly responsible for the persistence of the pollution. The dynamics of buried oil layers in the beaches at depths of some 50 to 80 centimeters have been extensively studied in several areas.

Pollution of the atmosphere

Tentative estimates of the quantity of oil evaporated into the atmosphere have been proposed, from a maximum of 90,000 tons, based on the chemical composition of the oil, to 60,000 tons comprising some 40,000 tons of light aromatics. However, the disappearance of the light hydrocarbons more volatile than n-C 12/ n-C 13 can be related to evaporation in the atmosphere and to their dissolution in the sea water. As far as human health is concerned, it has been shown that the content of toluene, benzene and alkyl derivates in polluted areas were smaller than in the center of the town of Brest.

Pollution of the sea water

During the first three weeks, the concentration of hydrocarbons in the sea was between 20 to 150 micrograms per liter. The concentration of oil in the water column was nearly the same whatever the depth, clearly demonstrating the importance of the vertical diffusion from the surface to the bottom. The period of decontamination (period of time during which a given concentration is reduced to one half) has been estimated to be 11 days in the offshore areas, 28 days in the bays of Morlaix and Lannion and 40 days within the Aber Wrac'h. By the end of July 1978, the pollution level was between 1 to 3 micrograms per liter, which is near the detectability of the analytical methods used and has no significance in terms of pollution.

Pollution of subtidal sediments

Two areas have been heavily polluted, the two abers, and the Morlaix and Lannion bays, and their decontamination cycles were followed by several laboratories. The pollution of the subtidal sediments can be directly related to the vertical diffusion of the oil in the sea. Nearshore, the local hydrodynamic conditions have also played a role, for instance in the bay of Morlaix. The decontamination process depends upon two main parameters, the sediment and the more or less protected character of the area. In April, 1978, the concentrations of oil decreased from more than 10,000 ppm (10 grams of oil per kilogram of sediment) in the upper muddy parts of the abers to 300 ppm in the muddy sand of bays of Morlaix and Lannion, and some 600 ppm in coarse gravel to fine sand in the eastern area (Primel). One year after the spill, the pollution level remained the same in the abers and was reduced to 150 ppm in the bays of Morlaix and Lannion and to 20 ppm in the eastern area. Important changes in the distribution of the sediment pollution in the bays of Lannion and Morlaix between 1978 and 1979 can be related to the effects of waves and currents on suspension and redeposition of oil-mineral fine particles during winter storms.

Weathering of the oil

After the physical processes of evaporation and dissolution of the light fractions, the oil had undergone a chemical evolution correlated with the energy level of the area, the oxidative reactions and the biodegradation. Briefly, the evolution of the saturated hydrocarbons is characterized by the fast reduction of the linear alkanes (the ratio n-alkanes/isoprenoids decreases), followed by an alteration of the isoprenoids. At the end, pentacyclic triterpens persist together with a mixture of napthenic compounds. The evolution of the aromatic fraction comprises a disappearance of the napthalene compounds, the phenanthrene and dibenzothiophene alkyl-derivates remaining with a complex of naptheno-aromatics.

Biodegradation

Biodegradation has been observed in the water column during the first weeks; low concentrations of dissolved oxygen, of nutrients such as phosphorus and nitrogen, combined with normal chlorophyll and silicium contents, can be explained by bacterial activity. The corresponding degraded oil has been evaluated to 0.3–0.4 mg/liter at the surface, decreasing with depth. The degradation rate is estimated to be 10 g/m³/ year at 10°C similar to previous laboratory studies. The total amount of oil biologically degraded in the sea could be nearly 10,000 tons, 5% of the total cargo.

   In several types of sediments, the microflora has been followed and the biodegradation estimated. Also, the chemical paths of the degradation have been studied; n-alkanes are degraded first, then isoprenoids, and then aromatics with 2 to 4 benzenic nuclei. The rate of biodegradation has been calculated for the intertidal sediments to be 0.5 kg/hectare/day. Assuming a total polluted area of 320 kilometers long and 0.5 kilometer broad, the biodegradation during the first weeks following the wreck should have been something like 8 tons/day. This estimation must be considered as speculative since it is an extrapolation of several analytical data. Specific studies have been conducted on sulfate reduction and methane production in marine marsh sediments.

Contamination of marine organisms

Most marine organisms accumulate the hydrocarbons, whether from the sea water, the sediments, or from their food. The behavior of some animals such as fishes and large crustaceans enabled them to escape from polluted areas. All studies with marine invertebrates show that the accumulation process of hydrocarbons is not selective, and mainly depends upon the oil concentration in the environment. As a consequence, the spectrum of hydrocarbons in a marine organism depends upon the weathering of the oil. For this reason, the isoprenoids and the polyaromatic compounds are the best indicators of oil pollution. The sites of accumulation in the tissues are different with the species: marine invertebrates accumulate the oil in tissues rich in lipids (gonads for instance); marine fishes accumulate in the gonads and gills. The decontamination of the marine organisms depends primarily on the level of pollution in the environment. Several studies have been conducted on cupped and flat oysters due to their economic importance in the abers and in the bay of Morlaix. Three main aspects have been investigated: the fate of the oysters in the polluted areas, the depuration processes of oysters removed from polluted areas to clean water, and the acclimatization of unpolluted oysters at different periods in polluted areas. Some technical comparisons were needed between scientists studying the global level of pollution and those working on the different hydrocarbons fractions and their dynamics, due to the fact that different methods and expression of results were currently used (spectro-fluorimetry, gas chromatography, hydrocarbon content in wet or dry weight). Unpolluted oysters have an average total hydrocarbon content of 5 to 10 ppm, dry weight, with an aromatic fraction below 2 ppm, dry weight. Oysters showing concentrations of 20 to 30 ppm, dry weight, for total hydrocarbons, are considered as slightly polluted by fossil fuels. On a practical basis, an average value of 60 ppm wet weight, was considered as the upper limit for human consumption.

The decontamination of the aliphatic compounds is a fast process in polluted areas while for aromatic compounds, it is essentially a slow and partial process. When removed to clean water areas, polluted oysters decontaminate in a few weeks for aliphatic compounds (from 90 to 4 ppm in less than a month); for the aromatic fraction, the rate of decontamination depends upon the time of exposure to the pollution. Histological studies of oysters' tissues demonstrated several cases of epithelial or gonadic necrosis which increased when oysters were removed.

Ecological effects of the pollution

Probably the first condition for a satisfactory assessment of ecological effect is to have a good reference or baseline qualitative and quantitative description of the communities exposed to pollution. This was fortunately the case in a large part of the polluted area, for intertidal and subtidal sediments, especially in the bays of Morlaix and Lannion, and the abers. Several commercially exploited species were affected including kelp, the edible large crustaceans (Dungeness crab and sea spider) and of course oysters.

Intertidal areas

The first main ecological result was observed during the progression phase, the end of March to the beginning of April, 1978. It can be called the "sharp mortality crisis" and has been previously reported in a very few cases, and on a very limited scale. The observations started on March 17, 1978, with some 250 field teams of four students each from the University of Brest. Rocky intertidal communities suffered variable losses; barnacles, mussels and *Sabel laria* did not suffer. The limpets mortality was about 30%. The periwinkles mortality was a little above 50%. Amphipods have almost disappeared, and isopods and decapods were much reduced.

On the beaches, very important mortalities were observed for irregular sea urchins, cockles, razor-clams, clams (Veneridae and Mactridae) and epipsammic crustaceans (mysids *Crangon*). A few kilometers from the wreck, all species have been killed, even those considered as resistant species (e.g., polychaetes and crabs). Sand macrofauna has also been killed under the low tide level: it was reported on the beach of St Michel en Grève (bay de Lannion), that some 25 million dead large invertebrates drifted up on the shore mainly from submersed sediments, together with a huge quantity of dead amphipods. This mortality is likely to be related with the high level of dissolved aromatics.

An interesting attempt to synthesize and quantify those field observations has been made using the average survival index of the herbivorous periwinkles as a reference. From direct densities and biomasse counts in both exposed and protected unpolluted areas, the survival index of periwinkles in polluted areas can be evaluated between 100 and 0 (no survival at all). This index, correlated with the quantity of oil coming on the beach, its chemical composition varying with the weathering process, and the number of times a given area has been polluted, characterizes the intensity of the oil pollution. This index is correlated with the survival rate of other species of the communities, using 9 species groups of differing vulnerability to the oil, from the more sensitive to the less sensitive (irregular sea urchins, cockles, amphipods and razor-clams, clams and periwinkles, *Scrobicularia* and *Lanice,* barnacles, polychaetes, mussels and seaweeds being proposed as leading species for each series of similar sensitivity). For each "commune" of the coast, a mean value of the periwinkles index is determined, and the corresponding survival values for the 9 series of organisms. Combining those figures with the standard biomass of each community and the corresponding area enabled calculation of the total loss in biomass as follows: for a surface of 200,000 hectares comprising 50% of rocky substrate, the loss in biomass would come to 260,000 tons in wet weight. This does not take in account the meiofauna (probably 10% in biomass of the large fauna), and groups of small crustaceans such as amphipods and isopods. Such a global assessment can be criticized for several reasons, the main one probably being that it assumes a constant mortality rate for a given species in a given area, which is not established. Still, it is also probably the first time that such an attempt was made on field data to evaluate the direct ecological impact of an oil pollution.

Detailed studies have been made on meiofauna, for the two dominant groups, nematodes and copepods; nematodes are the more resistant group, while the copepods showed changes related to the pollution.

In some polluted sandy or muddy beaches, the original community has been replaced by a new community comprising a very small number of tolerant species with a general composition similar to the opportunistic fauna of a sewage discharge area (polychaetes from the families capitellidae and cirratulidae). The recovery patterns of these communities follow more or less in time the ecological succession well-known around a permanent sewage discharge: apparently the oil, after the highly toxic phase, acts as an organic matter input.

Algae generally speaking are unaffected by the pollution. Only in the area of Portsall, at a distance of a few miles from the wreck, some species from the high intertidal areas have been destroyed drastically (*Pelvetia canaliculata* and *Fucus spiralis*), and the recolonization started one year after the wreck. At lower levels, an extension of the *Fucus vesiculosus* belt is correlated with the mortality suffered by the herbivorous gastropods.

Subtidal areas

In the bays of Morlaix and Lannion, the communities of the fine sediments, mainly fine sands, were heavily altered by the pollution during the first month following the wreck. The mortality was very severe for the amphipods, the irregular sea urchins and several gastropods. The amphipod *Ampelisca* represented by several species has been followed in detail: the three dominant species of *Ampelisca* almost disappeared in the summer of 1978. The total community of Pierre Noire in the bay of Morlaix (25 meters depth on fine sand) has been heavily reduced; density decreased to 20% and total biomass to 40%. The impact seems to have been more important in the bay of Lannion than in the bay of Morlaix. The restoration of the communities started in the fall of 1978. In the abers, the mortality was very severe, and the recolonisation starting in the spring of 1979 depends upon the remaining pollution level. In heavily polluted areas, low community diversity is typical with polychaetes (cirratulids and capitellids) dominant.

Some observations on plankton changes in biomass and composition showed positive recovery a few months after the wreck, and not too far from the coast. However, within the abers and the Morlaix bay, some modification of the normal planktonic assemblage was still present one year after the wreck, due to the persistent pollution coming from the large amounts of oil trapped in the muddy sediments.

Two years after the wreck, the major ecological problem to be followed was the restoration pattern of the different communities affected, in relation with the decrease of the pollution level. There is no question that on a quantitative basis the impacted communities at that time have not yet recovered their previous richness and diversity; however, on a qualitative basis, no species had disappeared completely from the total area polluted by the oil-spill. Among the problems not well understood, is the influence of benthic invertebrates of subtidal areas on the marine food chains. The amphipods, for instance, are by far the main prey for the young flat-fishes such as sole, plaice or turbot. The relation between the drastic mortality of the amphipods and survival and growth of the flat-fishes is unknown quantitatively.

Exploited species

The marine resources of this area are commercially exploited in four major ways: oyster culture, kelp and red algae harvest, and crustacean and finfish fishing.

The production of oysters is partly located in the abers, and most of it in the bay of Morlaix and the Bay of Lannion. The oyster culture technique is rather simple. The young oysters are deposited on the sea bottom and grow on the oyster flat until their size is appropriate for harvesting. This means that the oyster beds are always located in areas well protected from wave currents and wind action, i.e. in low energy areas. The relation is evident when looking at the concentration of oil within the oyster beds: all the oyster beds have been heavily contaminated, and so were the animals on the sea bottom. However, the oyster is a rather resistant invertebrate, and only very few animals were killed in the abers due to the very high oil content in the water and in the sediments. Most of them continued to grow, and remained contaminated until the end of 1978. Successive decisions were taken, leading to the destruction of about 6,000 tons of contaminated oysters. The abers were not used for oyster culture in 1979, although this now seems possible. In the bays of Morlaix and Lannion, oyster beds have been used for culture since the early summer of 1979.

In the impacted area, two groups of algae are exploited: the brown algae (3,000 tons of dry weight per year), mainly the kelp *Laminaria digitata,* and the red algae (1,000 tons of fresh weight per year), with one important species, *Chondrus crispus.* The effect of the oil-spill differs from one group to the other. The kelp belt which occurs just below the lowest tide level has not been affected by the oil. The reproduction of the kelp in 1978 occurred normally, and the density of young algae in the winter of 1979 showed usual values. The *Chondrus* population was affected by the pollution, as indicated by a general decrease in biomass in polluted areas for 1979.

The two main fisheries of the area are the crustacean fishery and the coastal finfish fishery. Crustaceans harvested include the Dungeness crab, the sea spider and a very few lobsters. The average yearly production is about 12,000 tons. No evidence of adult mortality was obtained in the first week after the wreck. Chemical analysis showed preferential hydrocarbon accumulation in the hepatopancreas (290 ppm) rather than in the flesh (40–60 ppm). The fishery statistics do not show significant differences between 1978 and 1979. However, the number of egg-carrying females was unusually low for lobsters in 1978 and 1979. This could have some consequences on the corresponding recruitment level, 4 to 5 years later. Something similar is suspected in the case of the sea-spider *Maja squinado.*

In this assessment of the ecological impact on exploited species direct evaluations are possible for the adults, but we cannot make good quantitative evaluation of a loss or a gain in the eggs and larvae. One can only make indirect estimations, that is to follow over the next year or two what is the strength of the year class born in 1978. As an example, the edible crustacean in this area, the "sea spider", has a production of about 8,000 tons per year, with a span of life probably no more than 3 years. Assuming that the totality of the eggs and larvae was killed by oil pollution in a given area in 1978, one third of the reproductive potential of this species would have been destroyed. In the case of the Dungeness crab, with a life-span of 12–15 years, the same consideration shows that only 10–12% of the potential of reproduction of this species would have been destroyed, which is probably within the range of the natural fluctuations due to other causes than pollution.

The effect of the pollution on marine exploited fishes has been extensively studied for flat-fishes, mainly sole and plaice. The flat-fishes have shown several types of fin diseases including fin erosion and blood irrigation in great excess since the summer of 1978. Fin alteration has been more severe in the abers than in the bays of Morlaix and Lannion. By the end of 1978, it was shown for the plaice that the juveniles from the 1977 year-class had a slow growth, that the three year old fishes were absent from the coast, and that no animals from the 1978 year-class were caught. In the abers, some alterations of the gonadic tissues, similar to those observed in the eel in the Roscoff area, were demonstrated by histological studies. Generally speaking, the flat-fish reproductive physiological processes have been affected by the oil pollution (non lethal effects) as well as the survival of larvae and juveniles. It is estimated that the population will be restored by 1982, when the year-class born in 1979 will be fully recruited.

CONCLUSION

One must underline the astonishing strength of nature to recover from such a huge pollutional impact as the *Amoco Cadiz* oil spill, at least in areas where waves, currents, and wind energy can freely work. However, in those areas more or less protected from the physical energy of the sea, the oil is still there, and the decontamination relies mainly on microbiological degradative processes which, in cold temperate environments work relatively slow. The complexity of the ecological impact was well demonstrated. To the direct loss in biomass and the corresponding one in terms of yearly production, one must add the sublethal long term effect, especially on the reproductive physiological processes, and as the food-chains alterations both of which are poorly understood.

The research program is underway for a third year. Unfortunately, a new tanker accident in March, 1980, spilling Arabian light fuel oil no. 2 with the same finger print as the oil of *Amoco Cadiz* will make the research more difficult, especially in the complicated field of quantitative estimates of the pollution effect on populations of exploited or non-exploited species. However, it seems probable that during the previous two years of study of the *Amoco Cadiz* case we have gained some 80% of all the results one can expect from such a program.

References

CNEXO, 1978. Premières observations sur la pollution par les hydrocarbures. CNEXO, Actes de Colloque no. 6.

CNEXO/COB, 1979. *Amoco Cadiz:* conséquences d'une pollution accidentelle par les hydrocarbures. Analyse bibliographique. COB, Brest, Novembre 1979.

Laubier, L. 1978. The *Amoco Cadiz* oil spill-lines of study and early observations. *Mar. Poll. Bull., 9 (11)*: 285–287.

Marchand, M., G. Conan et L. D'ozouville, 1979. Bilan écologique de la pollution de l'*Amoco Cadiz*. CNEXO, Rapp. Scient. Techn. no. 40.

NOAA/EPA, 1978. *The* Amoco Cadiz *oil spill. A preliminary special report.* W. Hess. ed.

CHAPTER 6

Economic Impacts

PHILIP SORENSEN
Florida State University
United States

This chapter represents the work of thirteen investigators involved in a study, which began in 1978, to determine the economic costs of the largest tanker oil spill in history. The study was coordinated through the Center for Ocean Management Studies at the University of Rhode Island and was sponsored by NOAA. The project had four components. I was the director of a component studying the market-valued social costs, which was concerned with fisheries and other damages where market prices can be used to determine the cost of the damage. A second study was concerned with non-market damages and in particular, tourism. The third part was the study of the clean-up costs and was conducted by two investigators from the NOAA agency itself in co-operation with French colleagues. The fourth component of the study, conducted by Thomas Grigalunas, studied the regional costs of the oil spill.

Our study group included various Americans, Canadians, and French investigators representing several different institutions. It's a rather extraordinary project since it involved American economists directing research in France, a country which could have decided that its interest in the major oil spill law suits, growing out of the *Amoco Cadiz,* might be compromised by the study. Nevertheless, we were able to go forward with the work. For my part, I found the French government officials, the fishermen, the many business people that I interviewed to be cooperative and interested. And in fairness, I think this is due to both the great diplomatic work of the NOAA people who helped us and our French colleagues, who facilitated our contact with a variety of difficult information sources.

We attempted to measure what I've called the social costs of the oil spill. The term social cost means the reduction in real income, including both tangible and intangible goods and services. We measured it from three different vantage points: the whole world, which would include the lost ship and its cargo, France, and the Brittany region alone. We were also very interested in the distribution of these costs from the Brittany region to the whole society. I will present a few highlights that represent the preliminary results of the study.

This oil spill was terribly serious because of its size. It was twice as great as the *Torrey Canyon* spill in terms of the volume of oil impacting the environment, and 20 times the size of the Santa Barbara oil spill. The spill affected numerous marine-based industries including tourism, fisheries, kelp harvesting, and oyster production. Of all the harvested seaweed in France 75% comes from this area, and the fishery and oyster culture industries (10% of the total oyster production for France) are of significant economic importance to Brittany. There also was physical damage to sea walls, piers, roads, and personal property. This was because of the extraordinary meteorological conditions at the time of the spill. The high winds and waves caused oil to be carried by the air onto crops and houses.

A lot of Bretons were put out of work for up to two months, and many Brittany industries saw their sales go down, because an image of oil contamination was created by sensational news stories all around Europe. Even the sale of uncontaminated oysters was affected by the suspicion that they were harvested in Brittany.

Finally, the psychological impact of the oil spill on the people of Brittany was enormous. The environmental insult resulting from the dumping of this terrible mess all over their beautiful beaches is part of the economic costs of the spill, but we will perhaps never get a true measure of that kind of psychological cost.

The fishing industry of North Brittany is a typical artisanal fishery. That is, the operation consists of very small boats owned by individual fishermen. There are about 700 of these boats in the oil spill area, and about 1,000 fishermen work on the boats. The species that are caught are primarily mackerel, bass, plaice, sole, pollack, lobsters, crabs, and several types of shellfish. The scallops from that area are regarded as about the best in France. These boats fish for different species in different seasons of the year, so this is a very hard fishery to handle in an economic model. It isn't a single fishery, where there is a single effort variable; it's a multi-species fishery.

Furthermore, the data for this fishery are scanty and they don't go back very far. This is because the Brittany fishermen are extremely independent. For years they wouldn't tell the government anything about their catch, and 1971 is the first year for which there is catch and value data.

There is also not any real management effort in this fishery; there is a licensing requirement, but the data is not compiled in any easily used format. A lot of our work involved counting the numbers and getting the data set up for analysis.

The 1977 pre-oil spill level of production in all of the open-sea fisheries, which does not include oysters, was 10,000 metric tons valued at 60 million francs in the oil spill area. The minute the oil spill occurred, the fishermen were closed out of the fishery, from March 17 to about the end of May. The fish catch for the whole region was down about 50% for the months of March and April, as compared to the year before. We modeled this in the form of a short-run forecasting model to see what the catch should have been in the period after the oil spill, as compared to what it actually was. The preliminary conclusion was that the French fishing industry there lost about 20 million francs in the ex-vessel value of fish and shellfish.

Oyster culturing is also important in Brittany. Very few of them (less than 5%) come from natural oyster beds. They are mostly produced in artificial environments using the well-known techniques of oyster breeding and culturing.

The two estuaries, Aber Benoit and Aber Wrach, were most heavily damaged. There's a lot of oil in there still as well as in the Bay of Morlaix, and these are three primary areas in which large quantities of the oysters are cultured. A lot of oysters were killed and many of them were tainted so that they simply couldn't be sold. There were exceptional costs in the attempted moving of oysters to clean water as well as the physical damage to the oyster parks and the sediments. About 6,000 tons of oysters were destroyed. The oysters have come back to Morlaix Bay, and they are being introduced back into the abers, but it probably will be several years before normal cohort and production levels are reached in the abers.

As I mentioned, an unfortunate aspect of the oil spill is that the oysters are affected by bad publicity now, and a lot of traditional customers are not anxious to look for the Brittany oysters. The diminishing of this effect will likely follow a degradation curve, and it will last for a while.

Neglecting the cost of cleaning up the parks or the sediments, the loss to the oyster producers was about 100 million francs or about $25 million in the study's preliminary estimate.

The seaweed harvesting industry is important in the sense that 75% of the seaweed produced in France comes from this area. In economic value, however, it isn't that important, and was only worth about a million dollars in economic value per year prior to the oil spill. So we're talking about a moderately important industry, but still important in that it produces a lot of valuable chemicals. This industry has been converted from a cottage industry to a highly mechanized industry in the last few years; from small boats with the fishermen actually picking the seaweed up with some kind of tool, to huge kelp-cutting boats which cost 700,000 francs. So the industry has been gearing up to harvest more and more of the total seaweed in the area.

The interesting fact is that the oil spill seems not to have affected the production or the harvest of seaweed. It is true that in certain areas you see seaweed growth lower than it was before the spill, but the harvest in 1978 was the largest in history. This includes all the adjustments for the new boats that came into the fishery in the seaweed harvesting area, and the necessary conversions from wet weight to dry weight since a lot of seaweed in 1978 was harvested and dumped into the factory fresh, rather than dry. In 1979, they actually put a quota on the harvest because too much seaweed was being harvested.

We had a lot of other components in the market-valued section of the costs. The ferry service that comes out of Roscoff was damaged because people from England just didn't come over. There are the wholesalers which deal in live lobsters and live shellfish and sell them for enormously high prices around Europe; they were put completely out of business. There are a lot of hot water health clinics in the area, and their business was down. If you take all the different components of the market value damages to France mentioned above, the total cost is around 140 million francs, which is very large by the standard of any oil spill that I've dealt with in the past.

The clean-up costs have been very high because of the volume of the oil involved and the places that it contaminated due to high winds and waves. It was estimated that 80,000 tons of the oil got into the sand and onto the rocks. Some 20,000 tons of oil were actually picked up off the beaches and put in ponds and finally ended up back in the refinery. The beaches were literally scraped and 7,000 men, including many French soldiers, were on those beaches picking up the seaweed manually, putting it in plastic bags and so on. So, the physical effort in cleaning up was really quite enormous and mostly involved military or public employees. But the clean-up is far from complete. Even now, the floor of contaminated estuaries are being plowed to try and expose the substrate to more oxygen to aid biodegradation of the oil.

It almost always will be true, that clean up costs are the most important component of an oil spill's economic costs, and that's true for the *Amoco Cadiz* as well. It isn't a matter of just adding up a number of vouchers to determine what clean-up costs were. For example, the army is not paid in France, yet there were 350,000 mandays of army time working in the cleanup. There were also 35,000 days of volunteer labor involved in the clean up. The investigators in this section, therefore, had to decide what kind of an economic value to credit to the army and the volunteers. The general conclusion was that the minimum wage in France might be an appropriate proxy for that.

If you add all of the costs of clean up, including these implicit costs, the preliminary estimate is about 350 million francs. This compares to a clean up cost of 80 million francs for the *Torrey Canyon* and ten and a half million dollars for the Santa Barbara oil spill. It is indeed, the largest clean up effort in history.

The evaluation of non-market damages really involves the loss to tourists. If you're a resident, you lost the opportunity to recreate in an unpolluted environment. There are survey methods which will allow you to ask questions from which you can determine a person's willingness to pay for an unpolluted environment. This is a way of putting a value on a clean site as opposed to the dirty site. The travel costs of people who come from outside the region allow the construction of a demand curve for a recreation site. The investigators in this section of the study used some extremely sophisticated economic and econometric techniques to look at the losses to the tourist. I'm only going to report some qualitative findings from the survey, and some statements by the surveyed population that I thought were most interesting.

On the whole, the tourists who usually come to Brittany expressed an opinion that the *Amoco Cadiz* oil spill was a significant disaster. They said it was worse than the student uprisings in France of May 1968 and it was viewed as only a little better than the construction of a nuclear power plant in their region. However, they felt it was primarily the residents of the area, rather than they themselves, who had been harmed.

The residents of the area rated the *Amoco Cadiz* as slightly less serious than the other crises rated by the tourists. So, apparently the residents who see the oil and live with it, are not as alarmed as the tourists. Most of the residents felt the media had paid too much attention to the spill and had exaggerated the extent of the environmental damage. In particular, they thought the appearance of the coast and the condition of the beaches had been exaggerated.

One third of the residents who were surveyed said that they had lost income because of the spill. Of those who said this, the income loss, on an average, was 27%. If you multiply the 2, you get about a 9% loss in income for the region, but an adjustment must be made since more surveying was done close to the coast than further back from the coast.

With regard to the quality of life, one quarter of the people in the area reported negative effects, that is, they couldn't go to the beach or they couldn't go fishing.

Three different surveys were done which were very expensive and elaborate. They were done by French survey research firms and cost a lot more than they would have in this country. You have to go through the survey several times to get the wording right. These are the types of questions that haven't been asked in France before; I'm sure you get very interesting answers, although some of them are not usable.

Regardless, it is clear that there is a sense of being very disturbed by the *Amoco Cadiz* oil spill. There is the feeling that it had significant adverse effects, but there is little in the way of actual behavioral change. When people were asked how many fewer times they went to the beach or how much they changed their activity, there was very little change shown.

There is the possibility that the appearance of damage to these people exceeds that which was actually suffered. As mentioned earlier, the perceived seriousness of the oil spill was least for the residents closest to the spill, moderate for the tourists within Brittany, but not close to the spill, and greatest for the tourists who didn't even go to the area, but were on clean beaches elsewhere.

Perceived damage seems to be related, then, to an individual's access to information and this led the team, who did that part of the study, to recommend additional study of the process of communicating information about an oil spill. I think that's not new for the *Amoco Cadiz*, but was also true for the *Argo Merchant*. Some studies indicated that people who never saw the oil from the *Argo Merchant* felt terribly damaged by it and were talking about all these fish that were dead on the beaches and so on, when none of that ever happened.

Dr. Grigalunas performed the regional part of the study, in which the region of Brittany was looked at alone. The social costs within the region were estimated, with the idea of developing policies that would compensate a region or give arguments for certain types of extraordinary policies to mitigate damages within a region. A lot has been said about compensating those regions threatened by oil pollution, by federal payments or transfers of one kind or another, since they do put up with a lot more of the direct costs.

The model used to estimate losses in tourism is an econometric forecasting model. It uses various independent variables to predict what income in the region's tourism businesses would have been in the absence of the oil spill. It thus looks at tourism-related industries to see how much actual income is below the expected income.

There are other categories of cost in the regional analysis, such as the fiscal costs to the local governments, the regional cleanup costs and so on. Looking only at the loss of income within the region of Brittany, it looks like the total amount was in the neighborhood of 200–400 million francs. I think it's fair to say that the size of the losses in Brittany, alone, are about half or a little more than half of the losses to France in total.

In summary, we can say on the basis of this preliminary work, that the *Amoco Cadiz* was, indeed, the largest in economic costs of any oil spill in history. Its total costs to France exceeded 500 million francs. That's over 120 million dollars. That is a lot less than the 2 billion dollars that I've seen in the press, but 120 million dollars in economic damage is a very significant number.

Thus far I have outlined the various categories of costs borne by France and by the Brittany region. As noted earlier, we also estimated the costs of the *Amoco Cadiz* oil spill to the world as a whole. Many of the economic costs suffered by France are costs to the world. There are, in addition, other costs such as the lost tanker and oil and some legal and research costs, which have been suffered by the world but are not costs to France. While we are not able to provide a detailed estimate of world costs at the present time, we anticipate that the total social costs of the *Amoco Cadiz* oil spill to the world will exceed 650 million francs or $160 million.* The effects on the marine environment, not cost assessed by us, were also enormous. So efforts to control or avert oil spills through improved crew or equipment requirements and redundancies seem overdue.

The people of Brittany have, for generations, taught their children that they were born of the sea. And they've seen this source of their life assaulted again and again in recent years by oil coming from foreign tankers, having no connection to them. More recently the *Tanio,* which was mentioned in Chapter 5, belched up oil onto the very shores in Brittany that suffered from the *Amoco Cadiz* oil. It's no wonder that the people of Brittany regarded our survey questions with a certain Gaelic fatalism. That is, they believed in their hearts that people really didn't care and that little was being done to help them.

---

*Final estimates of the costs of the spill are $145 to $175 million for France as a whole and $183 to $202 million for the world. The study methodology and results are described in *The Economic Cost of an Oil Spill: A Case Study of the "Amoco Cadiz,"* by Thomas A. Grigalunas and Joel B. Dirlam. (South Hadley, Mass.: J. F. Bergin Publishers, Inc., 1982).

CHAPTER 7

Policy Implications
VIKTOR SEBEK
Advisory Committee on Oil Pollution
England

While the general subject of this book is the impact of marine pollution on society, I will discuss, in this chapter, the other side of this process, namely the political currents within society which determine the development of the legal and technical rules of marine pollution.

In order to be able to set the policy implications of the *Amoco Cadiz* in historic perspective, it might be useful to examine very briefly the way in which international law governing marine pollution developed prior to 1978. This development shows a pattern which is very difficult to justify, both in scientific and in legal terms, although one can easily understand the political rational.

It is always easier to deal with easily packageable goods. For example, oil traditionally has received far more attention in international legislation than other, more toxic, substances. Most international conventions on hydrocarbon discharges regulate pollution from vessels, which it is generally agreed, present a less serious problem than the pollution from land-based sources. And yet, the very first important international legal instrument in this field, and it is a regional one in character, is the 1974 Paris Convention which entered into force only recently.

The reason for the haphazard pattern of the development of international law concerning marine pollution is that the legal machinery for adoption of international conventions at IMCO, for example, or regionally within the framework of European communities, is set in motion only when the sufficient degree of political good will can be found in most of the participating countries. That good will, as experience shows us, is generated almost over night after each major tanker disaster.

As a rule, technical and legal provisions which are adopted after such disasters only fill a particular vacuum, but do not necessarily improve the overall system in the long run. This results in a process which was described by the chairman of my advisory committee on oil pollution, as practicing "government by catastrophe."

It is thus to the *Torrey Canyon* oil spill that we owe the 1969 Intervention Convention, which allowed a coastal state to do certain things on the high seas in the case of a grave and eminent danger to its coast. It is also to the *Torrey Canyon* that we owe the two major international conventions on liability and compensation for oil pollution damage; the 1969 Civil Liability Convention and 1971 Fund Convention. The tightening up of the 1973 MARPOL and 1974 Safety of Life at Sea Conventions, even before they entered into force, is due mainly to a series of accidents off the United States coast particularly the *Argo Merchant* incident.

It is, therefore, not surprising that the *Amoco Cadiz* also generated considerable political activity and resulted in major changes or proposal for changes at IMCO and the United Nations Conference on the Law of the Sea, to which Dr. Waldichuk also referred in chapter 3. The *Amoco Cadiz* oil spill also spurred proposals for bilateral arrangements, which are now being set up in North-Western Europe, and in national legislation, especially in Britain and in France.

I think that perhaps one could also mention various voluntary schemes which are now proposed by industry, especially the recent amendments to Lloyd's Open Form Salvage Contract. The reason why the shipping and oil industry has gotten particularly interested in doing a certain amount of work, was not only for public relations purposes, but also because they knew that unless they came up with certain changes fairly rapidly, perhaps more stringent rules would be adopted by intergovernmental organizations, IMCO or perhaps the Law of the Sea Conference.

The 1958 Geneva Convention on Law of the Sea, produced four conventions primarily codifying existing customary international law. However, certain issues remain unresolved after Geneva, and, of course, the users of the seas and the relevant problems have changed significantly over the last twenty years; thus the United Nations decided in the late 1960s to develop further this very important area of public international law. Preliminary discussions started at a U.N. Sea Bed Committee in the late 1960s, and the Law of the Sea Conference opened with a very substantative session in Caracas in 1974.

The topic of this chapter was dealt with by committee three of the United Nations Conference on the Law of the Sea (UNCLOS), which focused on marine pollution, scientific research and transfer of technology. By 1978 most articles of the UNCLOS negotiating text relating to marine pollution had been decided. They reflected a fairly careful balance between the interests of flag states and also port and coastal states, although a number of states couldn't quite decide if their primary interest was as a coastal state or as flag state. Like every compromise solution, the package on environmental protection contained important shortcomings. In Britain, for example, it was assumed for hundreds of years that what is good for shipping industry is also good for the country. I think this was by and large true, until pollution started presenting such a serious problem to an island state.

The *Amoco Cadiz* provided an opportunity for France and some other states to propose improvements in the negotiating text at the Law of the Sea Conference. As a result of this, the following provisions were changed and acquired the status of proposals which received the general consensus. Article 221 now allowed detention of vessels for pollution offenses committed in the exclusive economic zone, not only in the territorial sea. Moreover, a physical inspection of the vessel was allowed, provided clear grounds existed for believing that a vessel had violated international rules and standards and that it navigated either in the territorial sea or the exclusive economic zone of that coastal state. The violation had to result "in a substantial discharge, causing or threatening significant pollution of the marine environment." That word, "threatening," was the key change that was made in the article after the French proposal.

Article 222 was also amended to provide that nothing in that part of the convention would prejudice the right of states to adopt and enforce measures beyond the territorial sea, proportionate to the actual or threatened damage, to protect their coast line and related interests (including fishing) from pollution or threat of pollution following a maritime casualty.

Article 231 provided that coastal states were no longer limited to imposing only monetary penalties for pollution offenses. This was something which some states had proposed at the 1973 Conference, but most maritime countries, including Britain, were very much against it.

Finally, in 1979, it was also agreed to adopt an article which would permit states to cooperate in developing international law on the responsibility and liability of states for compensation or other relief with respect to principles to be elaborated at the Law of the Sea Conference. They argued that the 1910 Brussels Convention on the unification of certain rules and laws relating to assistance and salvage at sea was now inadequate because that convention treated salvage as one of a private contractual relation between salvor and the ship, while the pollution over the last ten or twenty years had introduced an important public law aspect. In particular, France argued that the state ought to be allowed to request, if the need arises, compulsory salvage. This, of course, is contrary to the very concept of salvage as defined in the 1910 convention whereby nothing could compel a salvage company to intervene and where, if salvage operation was unsuccessful, no payment might be due.

The situation at the present is that the Committee Maritime Internationale suggested that they would be willing to help with private law aspects of a possible new convention, while the legal committee of IMCO would concentrate on the public law aspects. However, before the *Amoco Cadiz* oil spill it was decided that on the basis of preparatory work to be carried on by the legal committee, a new diplomatic conference would be convened in 1982 to adopt the first convention on compensation for pollution from chemicals. Most efforts of the legal committee concentrated on this topic, and no significant progress has been made so far on discussion of salvage. The same applies to the right of coastal states to intervene in the adjacent waters, although France was fairly successful in persuading the Law of the Sea Conference to change article 222.

It is probably fair to say that the issue which raised the greatest controversy was compensation for oil pollution damage. This, of course, is not a new area in international environmental law. After the *Amoco Cadiz* oil spill, two voluntary industry schemes were set up, and two IMCO-sponsored conventions were proposed, as noted earlier; but it was argued before the *Amoco Cadiz* incident that the 1969 and 1971 conventions, which have been in force only since 1975, were not really adequate to cope with the problems which are highlighted by the *Amoco Cadiz*. However, in view of the work on a convention for compensation from chemicals, no significant progress has been made so far on this issue. There is, in theory at least, a possibility that amendments to those two conventions might be adopted at a new diplomatic conference in 1982, if sufficient preparatory work is carried out before then.

What I propose to do is to identify a number of areas where changes might be necessary and where certain proposals have already been made by various states in the legal committee of IMCO. For example, it seems paradoxical that in an area where sovereign rights of coastal states are recognized, in the 200-mile exclusive fishing zone and exclusive economic zone, a regime for compensation for pollution damage does not extend. It stops at the limit of the territorial sea, which is up to twelve miles.

It has also generally been agreed that the current limits of compensation in 1965 and 1971 conventions are not adequate; but, unfortunately, if one attends the meetings of the legal committee, it reminds one almost of an oriental bazaar, or an auction. "Shall we have a limit of 50 million or 70 million or 100 million?" Such bargaining doesn't really relate the limit to any meaningful scientific or social parameters.

At present the limit in the 1969 convention is some nine and a half million pounds. While the first assembly of the fund convention raised the 36 million dollar limit to 54 million dollars, in theory, this could be raised further to 72 million dollars. That is an absolute maximum, however, beyond which the fund could only go if the convention were to be amended.

After the report which the French National Assembly drafted on the *Amoco Cadiz* oil spill, a lot of proposals were drafted at the legal committee, with a view to developing a special rule under which the insurance industry might help in promoting tanker safety and reducing pollution and also penalizing those operators who use substandard ships. Also, it was suggested that perhaps provisions should be made for payment of interest and making interim payments to victims of pollution, because delays in settlements are considerable.

In addition that tacit amendment procedure, which is already adopted in some IMCO conventions, should also be incorporated in liability conventions. At present, for example, if one wanted to make changes quickly, it just wouldn't be possible. It takes up to two or three years to prepare for a diplomatic conference, an average of five years for an IMCO convention to enter into force, and a further two or three years for it to be then incorporated into international legislation.

The oil and shipping industry have also argued that there might be a more careful balance between the responsibility of a carrier or shipper and cargo owners. This is provided in the 1969 convention; liability is provided in the 1971 convention.

It is argued, particularly by local governments through such forms as the International Union of Local Authorities, that compensation also ought to be provided for what you call here in the states "mystery spills," or unidentified pollution. At present, no compensation is due in international law.

It should also be mentioned that the French delegation raised the issue of flags of convenience. This is another emotional issue, which was not resolved, at the 1958 Geneva Conference. It wasn't enough to tell the flag states that there ought to be a genuine link between them and the ships they register. Another issue is to make sure that various states do not allow their physical and legal persons to register ships in flags of convenience states. The French delegation asked that, at least, the relationship between the ship master and ship owner ought to be examined. The only result of this request, however, was a non-binding resolution at the eleventh assembly of IMCO last year, which suggested to states that they ought to give stronger powers to ship masters. It was suggested that if a ship master already has considerable responsibility in running a ship, he ought not to be in a position to Telex or telephone his superiors in the other corner of the world to ask them what to do next in the event of an accident.

Substandard ships was another issue that the French delegation raised, suggesting that IMCO should ask the oil and shipping industries to provide information on what measures they propose to take to insure that the ships which they use are not substandard. This is a very important problem, because if you look at the statistical data, it is fairly simple to determine which countries have a fairly good safety and pollution record. It is also possible to establish which oil companies have a good record when operating their own vessels. However, there is very little information available on the record of ships chartered by major shipping and oil companies. Experience shows us that a lot of major incidents involved ships that were chartered to an otherwise perfectly respectable oil company.

It is rather interesting that after the *Amoco Cadiz* a number of major oil companies, some of them American based, stated that it is now part of the company policy that substandard ships would not be hired.

I conclude in discussing the activities of IMCO that the organization stated at the thirty-eighth session of the Maritime Safety Committee, that it was competent to deal with all marine safety and marine pollution issues and that efforts should not be dispersed. In plain language, this position was really an attempt to tell other international organizations to keep their fingers from the pie. This, I think was justified to a certain extent, because perhaps too many organizations are doing the same thing.

However, work, which the United Nations Environmental Program does in regional seas, for example, does not duplicate the work of IMCO. In fact, this work develops marine pollution protection further, and the Law of the Sea Conference has encouraged development of regional environmental law. I believe that the same could be said for such regional schemes as the 1969 Bonn Agreement, which was ratified almost a year after the *Amoco Cadiz* oil spill. The reason for such expeditious action was that the agreement didn't require parties to the agreement to do anything in particular, but merely to exchange information. However, after Ekofisk, it was agreed that the states should also deal with pollution from offshore operations. And after the *Amoco Cadiz* disaster, it was agreed that they should also look at pollution from chemicals.

Important work on marine pollution, particularly relevant to the question of compensation, is also being carried out by OECD and, in particular, by its working group Transference of Pollution. They have studied different aspects of compensation and the calculation of damages, cleanup, and other costs. I simply cannot overemphasize the importance of this issue, because, as far as compensation is concerned, we are dealing with a rapidly developing branch of international and environmental law, which is being developed in comparative obscurity, unlike other aspects of compensation or insurance, such as hull insurance or cargo insurance. These claims are settled fairly quickly, sometimes only in a matter of weeks, although multi-million dollar sums are involved.

Third party liability, that is, liability for all pollution, is being mainly dealt with by P and I clubs, which is a practice that dates back a few centuries. This approach applies where you have a certain sector of industry, which owns the insurance companies in which they are the assureds, and who are able to negotiate claims with victims of pollution. The victims sometimes may be under considerable pressure to settle quickly or else they might have to wait from two, to five years. For a small businessman such as a fisherman, unless his government will provide an interim payment, he simply might not be in a position to wait that long.

One international organization, which should concern itself with pollution, is the European community; and there, perhaps, we can particularly notice a significant impact which *Amoco Cadiz* had on society. The European community is an international body, with special powers, including powers of enforcement. IMCO, or member states under the aegis of IMCO, can adopt various conventions, but cannot be responsible for enforcement. Member states are responsible. However, the European community has powers of enforcement and member states can take those who do not comply with them to their special court.

It is rather interesting that after Ekofisk, the European Commission was anxious for the Council of Ministers to pass some broad resolution and involve the community in this area; but because it didn't become an important political issue (after all one doesn't win or lose an election over marine pollution) they decided that they would simply refer the matter for discussion at some further point.

However, after the *Amoco Cadiz*, the French government was successful in persuading the Council of Ministers to adopt its action program on prevention of marine pollution and, as a result of that, six or seven major studies in different technical and legal aspects of marine pollution were carried out. The European Commission is expected to make a decision on the studies before the end of this year, and one of them, in fact, analyzes the role of marine insurance in the system of compensation for pollution damage.

I would finally add that a lot of significant changes are required in those countries which are most immediately affected by *Amoco Cadiz*, particularly France and the United Kingdom. It is a case in many countries that the responsibility for marine pollution, as indeed other matters, is widely dispersed among all sorts of ministers. In the United Kingdom, for example, marine pollution is under the jurisdiction of the Department of Trade, of the Department of Environment, of the Minister of Defense, of the Foreign and Commonwealth Office, of the Department of Energy, of the Scottish Office, the Welsh Office, and so on.

Norway, I believe, was one of the first countries in the world to give a minister, without portfolio, a special responsibility for all maritime matters, but France became the first country actually to institutionalize the system of coordinating maritime policy, when in August of 1978, they set up an Inter-Ministerial Committee on the sea. Admittedly, it deals with all aspects of maritime policy, but particularly marine pollution because it was prompted by the *Amoco Cadiz*.

France also quickly passed several laws in 1978 and 1979, which increased the levels of penalty. Penalties in most countries are still ridiculously low and therefore could not possibly act as a deterrent. France also introduced special rules on reporting of accidents to the coastal state, something which France also raised at the Law of the Sea Conference. At present the 1973 MARPOL Convention merely requires a ship master to report a major spillage or a threatened spillage; France quite rightly argued that quite a lot of minor defaults in ships could, in fact, lead to a major catastrophe. They argued that a coastal state ought to be in a position to follow what goes on from the outset.

Now, what are the remaining loopholes in dealing with pollution? I think that where the implications of the *Amoco Cadiz* were considerable, a tangible success could only be claimed at what happened at the Law of the Sea Conference. At IMCO it is still not tangible, mainly because most of the issues at stake were under the jurisdiction of the legal committee and are now primarily dealing with chemicals. It is perhaps also expected, and hoped for by certain countries, that political pressure from France will eventually decrease. I don't think that this will happen, at least not until the *Amoco Cadiz* court proceedings in Chicago are terminated.

Perhaps one of the useful consequences of the *Amoco Cadiz* oil spill is that IMCO which played a leading role in the formulation of law on marine pollution and whose role is likely to be enhanced when the Law of the Sea Conference produces its final convention, has now accepted that enforcement of existing rules will be given the highest priority. Thus the Secretary General, Mr. Trivastara, made his very publicized statement at the eleventh assembly last year, that "There will be no new diplomatic conference in the next two or three years and that his organization will, in fact, concentrate on enforcing the existing ones."

There exist several international conventions, but quite a lot of them are still not in force. Some of the conventions which are in force, are already technically out-dated, and enforcement procedures are still in the hands of flag states, which even in the best cases very often simply cannot properly enforce their provisions.

Previously, I spoke critically of the short-sightedness of schemes for improvement on the law of marine pollution, which are based on the experience of one single accident. However, I think that it must be admitted that the threat of unilateral action could serve as the most important impetus for refusing to settle for the lowest common denominator and, in fact, to speed up the regulatory process. Therefore, paradoxically, any firm progress in the areas, which I outlined in this chapter, and also in other aspects of marine pollution, may well depend on the future incidents like the *Amoco Cadiz*.

CHAPTER 8

Commentary
Superfund Legislation –
What Will It Mean?

KHRISTINE HALL
U.S. Department of Justice
United States

My background is as the chairman of an inter-agency task force, which examined issues dealing with potential superfund legislation. I was handed that task the second day on the job at the Department of Justice and my remarks are going to be necessarily colored by the battles the administration and I have fought since 1979 in trying to come forward with a cogent piece of legislation and in trying to get it passed.

The administration's superfund proposal was an all encompassing proposal that would cover pollution, both marine and land-based, from both oil and chemicals. In early 1979 various agencies from the federal government made a number of policy decisions on how superfund should look with regard to liability and compensation for the releases of either oil or chemicals.

When you're dealing with liability and compensation, two real questions arise; first is the question of deterrence. This unit is titled *"Accidental Pollution"*, but as was pointed out earlier, accidents can be prevented. The *Amoco Cadiz* incident could have been prevented. How do you deter or prevent further accidents from happening in the future?

The second very important question that has to be asked is who, in our society, should bear the risk? Who should bear the burden when damage does occur? Should it be the victim, should it be the taxpayer, or should it be the consumer of products, indirectly through the industry, who bears the burden?

The developed countries have come down squarely on the principle of the polluter (industry) pays. The industry then, of course, passes its costs to the consumer. Industry must realize right off the bat that they're going to pay the costs. And I stress the full costs; not just a portion of the costs, through very low limits on liability, or just a portion of the costs, through not taking into account all of the various kinds of expenses which occur as results of a release. If industry recognizes that they are going to end up paying, that their liability is certain, and that they will not be able to hide behind a whole tangle of legal arguments, accidents will be prevented in the future. Also, a strong liability system focuses the risks of an activity more precisely on that segment of society which benefits more from the activity.

It's for these reasons that the administration's superfund bill proposed joint, several, and strict liability for releases of oil and hazardous substances. Needless to say, it's been the source of a lot of controversy in Congress. Because of the intense lobbying on liability, superfund may not be a strong piece of legislation.

After superfund legislation is passed in whatever form, we have the second, and probably more important, question of how that piece of legislation will work. If the polluter pays, what does he pay? There are obviously a whole range of costs to be considered including possible medical costs; Dr. Sorensen discussed in chapter 6, the psychological damages that occurred in the spill. This seems to be a recurring theme in some of the recent big incidents; psychological problems have cropped up at Love Canal and Three Mile Island. I was not terribly surprised that there were some psychological damages from the *Amoco Cadiz* incident.

Cleanup costs are only a portion of the actual costs of an incident; natural resources are often damaged. Who pays for replacement or the loss of natural resources? Various other economic costs include property damages, damages to commercial fisheries, loss of tax revenues, loss of tourism, and all the secondary effects.

If these costs are to be assessed against a polluter, how do you measure them accurately and without undue time or cost? It has taken, as far as I can see, several hundred thousand dollars just to establish what the costs are from an incident such as the *Amoco Cadiz.*

These are the sort of things that can really tie up the implementation of a piece of legislation such as superfund in the real world, once it gets passed. If the government, or a victim, is not able to measure his damages accurately on a timely basis, the deterrence and the risk spreading system begins to break down.

So, I'd like to issue a challenge to both the economists and the scientists who are working on trying to establish the costs of pollution. I congratulate them, first of all, for using the *Amoco Cadiz* incident as a case study to test some hypotheses and further the methodology, but we really have to develop all of this further if we're going to have a compensation and liability system that works. Sometimes when I get very frustrated about what's happening in Congress in terms of getting superfund passed, I remind myself that the battles are just beginning.

Editor's note: The final version of superfund is considerably less detailed than the earlier drafts which proposed a $4.1 billion fund from fees assessed on industry, and allowed compensation for personal injuries and gave injured parties the right to sue in Federal court. Basically, the legislation establishes a $1.6 billion fund, over 85% of which will be derived from taxes on industry, which will finance the clean up of hazardous substances released to the environment. The superfund legislation as passed is somewhat vaguely written and will invite legal challenge during its implementation.

Discussion

Comment: I agree with Miss Hall's last suggestions that the *Amoco Cadiz* incident was a preventable accident in both a technical and a legal sense. I think it was more tragically a legal failure than a technical one, in that salvage operation waited for hours before any action was taken. This was because the ship's master was on the telephone with the company agent in Chicago, who told him strictly not to sign the Lloyd's Salvage Agreement. I think that this legal problem really gave us the biological and economical problem we've been talking about.

I would like to add something to Dr. Sorensen's remark about economic costs. There was one big category omitted, perhaps intentionally. This is the ten million dollars or so of legal fees that are certain to be generated by the fifteen or more major legal actions filed in France and in the United States. If the figure of $100 million is correct, the 10% for legal fees isn't out of order and it could go much higher. It's something that probably should be added into the total economic picture. Since those costs will certainly come out of the top of any settlement that's reached.

Mr. Sebek has mentioned a number of changes that were made in the articles of the Law of the Sea. France was able to convince the other parties there that these were very good changes to make, increasing coastal state authority over flag states. Mr. Sebek failed to mention the reason they were so convincing was that the rest of the world community was quite worried that France would take unilateral actions beyond the initial immediate response of putting up a safety zone around France where no tankers could transit. There was also considerable fear that France, located in that critical transit area, would exert even more unilateral authority in that area.

I also think some mention should be made of the efforts of the International Salvage Union (ISU) to amend the Lloyd's no cure, no pay policy. That is clearly the trouble spot legally and the CMI working with Lloyd's and the ISU are making some real advances in this area that should be mentioned as well.

Sorensen: Extraordinary legal costs to the governments in France and other parties are part of the social costs of this spill and should be included, assuming that they are extraordinary or marginal legal costs for the world. Some people might argue that there is excess capacity in the legal profession, and that, therefore, the *Amoco Cadiz* spill will not require society to give up any useful alternative legal services. In fact, we did include extra-ordinary legal costs in the Santa Barbara oil spill damage assessment.

Grigalunas: In our accounting for the social costs of the *Amoco Cadiz* we do have a category which represents costs to the world and legal costs, as well as research costs. Getting some estimates of the extra legal costs that are incurred are extremely difficult, but I'm hoping that any lawyers who are here will provide some of that information.

Sebek: In this area one often avoids examining the morality of the success of any action. I would say, for example, that when President Carter made his famous statement in March, 1977, that "either IMCO will tighten up MARPOL '73 convention or we shall go ahead unilaterally," a lot of countries, including the United Kingdom and France said "We don't want to be told by the United States what we ought to be doing. You are really blackmailing us." Yet, this was the first time in the history of IMCO that within less than a year a major diplomatic conference was prepared and two major pieces of legislation were adopted: the 1978 protocols to MARPOL '73 and '78. Lots of civil servants in the United Kingdom, France, and Germany were unhappy with Carter's initiative while congratulating themselves for achieving a compromise in February, 1978.

Comment: I agree entirely with your conclusion that legislation by catastrophe is simply the way things happened in the past. You have to live with that and, I guess, not feel too badly about it, but I was trying to make sure it was understood that France did not all of a sudden become very persuasive. It was really their threat of unilateral actions that created the climate in which these changes were made possible, and certainly the same thing has been true with IMCO and their sudden ruling to discuss many issues including salvage, manning requirements, etc.

I'd like to ask a question about the psychiatric costs. It seems that this has been mentioned and that the media has exaggerated these costs. How are you going to get the media to pay for what they created?

Sebek: Censorship might be one way.

Grigalunas: One part of our *Amoco Cadiz* study involves an informal survey of German chartered tour operators. The German tourists represent one of the most significant shares of foreign tourists that visit Brittany, and a very large proportion of tours were cancelled or redirected from Brittany to other parts of France in large part because of the adverse publicity. Our understanding is that it was very much exaggerated and out of date information. The publicity itself, which may not square with the facts, generates perceptions which influence actions.

Comment: The media's coverage of the Three Mile Island incident was a comparable situation, but it seems difficult to charge the media for the costs they have created.

I'm of the impression that in the incident at Bedford Bay there were after effects discernable for five or more years, when anyone looked critically at sediment conditions and the transfer of contaminants and sediments into the biological community. The comments I heard earlier this afternoon was that the biology affected by the *Amoco Cadiz* may be returning to normal in reasonably short order. These comments seem over optimistic in relation to the real damage to the environment.

Laubier: I don't exactly understand what you are saying, regarding the consequences of the Bedford Bay accident, but you find what we said about the *Amoco Cadiz* to be a little optimistic. At the conclusion of my paper I stressed that one of the major impressions, two years after the accident, is the huge effort still required for natural forces to bring about recovery to the environmental conditions before March, 1978. The remaining problems are now restricted to the upper parts of two or three small estuaries in which high concentrations of oil are still present in the sediments. The more toxic parts (aromatics) in muddy sediments have disappeared, but concentrations of about one to ten grams of oil per kilogram of sediment are still measurable more than two years after the wreck.

The biological communities are for the most part, recovering, except for the upper parts of the abers and the marine marshes. I agree fully with you, and as with the experience of *Torrey Canyon,* it will probably take ten years before the ecosystems are recovered in the quantity and diversity of species. Still, looking at those 300 kilometers of coast polluted in April, 1978, the main conclusion remains that recovery is proceeding very fast.

Most of the polluted area is characterized by strong wave and current energy, playing a large role in pollution dispersion. Researchers have established a predictive index of the ability of the different parts of the coast to recover. Exposed rocky coasts may recover in six months to a year. Recovery may take ten years or more in the abers and marine marshes.

Comment: What did you mean by "no longer toxic?" You said that the "oil is there, but it's no longer toxic."

Laubier: The toxicity of the pollution decreases with time. In the abers sediment, all the aromatic products went away during the first two to four months by evaporation and other means. What remains is bound strongly with the sedimentary particles. In the abers, there are a lot of oyster flats. At the beginning the idea was to clean the abers as soon as possible. For several reasons it was decided to clean and remove some twenty centimeters of sediment from the oyster flats and adjacent areas. The upper part of the estuary, in which there are no oyster flats because of the low salinity, was not cleaned and remains the main source of pollution because oil is trapped in two or three meters of very soft sediment. And now, removing these sediments is not possible because the oysters downstream would receive pollution from the resuspended particles and be contaminated again. The oyster producers are just beginning to operate again and strongly oppose such suggestions.

Sebek: Legally, of course, the concept of compensation for pollution damage caused to the marine environment, per se, is relatively novel in international law. In fact, in the early years of the Tovalop and Crystal agreements, compensation really meant reimbursement of clean up costs, plus a little to fishermen and the tourist industry, because compensable damage had to be direct, and not remote. It was very much linked with the concept of property. For example, a federal state could not claim damage as a guardian of the environment. Compensation for damage to the environment, per se, was simply not recognized.

But this is a very fluid area. Lots of things are happening very quickly, even regarding psychological damage. It was only twenty, thirty, or forty years ago that it was fairly peculiar to claim damage, for example, for shock or pain suffered after an accident. We may well get used to exactly the same relation to damage to the marine environment.

Comment: In the matter of superfund, my understanding of it is that payments are made on a per barrel basis by the shippers, and then compensation payments are made from that created fund. If I'm correct, it seems that there's a big decoupling between the causing of an accident and the compensation for liability, and that deterrence is reduced at least. Also the distribution of the payments is among both those who do and don't cause accidents.

Hall: Under most of the early versions of superfund, an assessment would have been made on oil companies of a few cents per barrel and on chemical companies based on a unit of feed stock used in the production of that chemical. From whatever tax source, however, the money goes into a fund, and any money that the fund pays out (e.g., cleanup costs), the fund can get back through subrogation to the victim's rights and by going after the polluter for the costs. The initial stage has the big fund established on the basis of a contribution from everybody, regardless of fault and activity, but when you get reimbursement for individual incidents, you tie it back to the polluter and handle it case by case.

Comment: Is the fund responsible for negotiating to get that back?

Hall: Yes.

Comment: You don't continue to pay three cents a barrel?

Hall: That's right. In fact, there is a cut-off point after, I think, four or five years, at which time the fund administrators will have to come back and say either, "We have enough. We don't need any more. Or, please give us some more."

Grigalunas: The question really is an important one, because it addresses the issue of compensation and also the creation of an incentive system to encourage "proper" behavior on the part of polluters. If polluters are forced to bear the cost of a spill, then they will recognize the costs that might be borne and take actions accordingly. I'm not sure of the details of superfund, but it's very clear in the Offshore Oil Spill Pollution Fund, that the company that causes the pollution would be assessed the damages, and the transaction cost of bringing suits is really reduced.

Hall: The superfund is really the OCS scheme on a larger scale with some variations.

Comment: I wonder if Dr. Sorensen would say a little bit more about the "willingness to pay" studies that were done and how successful they have been. What kinds of ranges of numbers came out of them and how useful have they been?

Sorensen: Elizabeth Willman, with Resources for the Future, and Gardner Brown did the actual research and I prefer that she speak to the issue.

Willman: We tried a number of different approaches to get some estimates of what is called "willingness to pay," i.e., what is the worth of clean beaches in the Brittany area to consumers (tourists) of the services of these beaches? Residents were also considered in the study.

One of the studies that was used was an annual study carried out by an institute located in Rhen, that interviewed households in the area each year and collected some information on tourists such as how far they came and the distances they traveled to get to the beach locations. One of the things that we did was to assess the costs that people incur traveling to get to a site, in order to derive an estimate of how much that site is worth.

Another study that was done by a survey research firm in Paris: in this study people that were visiting beaches in 1979 were shown pictures of what the beaches were like when they were first dirtied in the early spring, and asked how much further they would be willing to travel to get to a beach that was cleaner by a certain amount. That did derive some estimates.

I can't at the moment use these estimates to give you an accurate estimate of the total costs because there are a few pieces of information still missing that relate to the proportions of the sample population that came from different areas. But to give a very preliminary number, the costs for a family might be in the range of $40 per season. That's probably a relatively high estimate.

Gardner Brown also did some work using a technique which examines whether the hours people spent visiting clean beaches could be determined to be worth more than equivalent hours spent at dirty beaches, using information on expenditures that people made. This was less successful and the cost estimates were smaller.

There are several qualitative costs in the study that were mentioned, but for which we do not have estimates. One of the things that came out of the study relates to psychological damages. Fishermen were asked if they lost any income due to the spill or was the quality of their life, in some way, made worse? In a number of cases the answer was no. However, they still felt very strongly that this had been a terrible disaster. That seems to me to be some indication that there is a reaction not just to the spill, but to the institutions that allow this sort of thing to happen.

Grigalunas: I think this is really somewhat of a technical area, but a very challenging one, especially from the point of view of the economics profession, because it represents an attempt to place economic value on an experience where there are no market data or only indirect market data. Yet the costs can be very substantial in the case of many spills.

Looking back at the study that Phil Sorensen and Walter Mead did on the Santa Barbara oil spill, the third highest cost of that spill was the subjective loss by recreationists, resulting from the spill. So, there can be substantial costs resulting from oil spills that are sort of non-market costs along the lines that Ms. Willman mentioned. They are extremely difficult to measure in the United States and even more difficult in another country, with culture and language difficulties.

Comment: To get back somewhat to the biological/economic problem, I wonder how carefully certain topics will be studied in the future. I refer to the possible future loses to the oyster industry that may occur, because of a possibility of the estuarine sediments serving as a toxic reservoir of materials, to be released over a fairly long period of time, and thereby affecting the population within the area. Is it possible to put an economic value on this?

Sorensen: This is a most difficult problem, but we do have to go to press sometime. We can't wait forever to determine what has happened to the oysters. But before we go to press, we will need to talk to the oyster producers in the abers and see how the situation is evolving. They have been in the business for years and they have pretty good judgment about how things are working out.

We will also be talking to people like Dr. Laubier, to find out what, in fact, the research is indicating. Ultimately, when we publish our findings next year, we will only be able to hypothesize about the long-run damage to the marine environment, and until a future economic study is undertaken, our results will have to stand.

Comment: Can any estimate be given as to the possible long-term effects on the oyster industry?

Laubier: The only areas still containing high concentrations of hydro-carbons are the abers. The long-term effect on the oyster industry of the abers is difficult to evaluate because the abers are used as growing flats as well as "finishing" areas to allow the oysters to develop the special taste from living in brackish waters for a few months. Over 500 tons of oysters are cultured per year in the two abers. It is difficult to know how many more oysters are placed in the abers for a short time, just for "finishing."

The rest of the area, the Bay of Morlaix and Lannion, is also a major production area, producing about five to six thousand tons per year. This area is now decontaminated. The major problem remains in the abers, but it's a risk due to the fact that in the upper parts, especially in the aber Benoit, (about two or three square kilometers of mud flats) may be polluted for ten to twenty years.

So there exists a risk if something happens to those muddy flats such as dredging or heavy rain; but I'm not even sure that natural conditions can affect them.

Studies have also demonstrated that oysters show some kind of disease within the gonads. However, the significance of this fact is controversial since, ordinarily, due to temperatures of the area, the oysters do not reproduce. Because the oyster seed usually does not come from this area but from the south of Brittany, the production is not limited even if the oysters in the polluted area do not produce normally.

Grigalunas: This is another related dimension on long term effects and it seems that these are the costs most difficult to measure and, perhaps, given a lesser amount of attention than they deserve. I suppose it's fair to indicate that all studies have a finite time horizon and you have to finish at some point. Perhaps we don't have the predictive tools and adequate ecological models that allow for an assessment of long term effects.

But it is interesting in a policy setting. In the OCS Lands Act, which includes the Offshore Oil Spill Pollution Fund, there's a time limit on liability for damages.

Comment: Miss Hall, the superfund will pay for a variety of kinds of damages. We've heard Mr. Sorensen talk about the kinds of damages they've been measuring on the *Amoco Cadiz* such as clean up costs, losses to fishermen, losses of recreational opportunities, and losses of income to people who run the hotels and restaurants, as well as lost recreational opportunities of those who travel to these areas. What kinds of things will the superfund pay for?

Hall: As I mentioned there were several versions of superfund on the Hill. The version of superfund which passed is limited. The House versions had other types of damages. The Senate version was probably the most expansive, in terms of what kind of damage it covers. Primarily, the legislation provides for clean up costs. The degree to which it can be applied for other damages will likely be the subject of litigation. Superfund may now be law, but how it will be implemented remains unknown.

Grigalunas: It's interesting that the one bill we do have on the books is reasonably comprehensive, but doesn't include certain categories of costs that we were measuring in the *Amoco Cadiz* project, such as the loss of recreational uses of the beaches. Of course, it's not easy to compensate individuals using beaches who suffer damages. But there are categories of cost like that that aren't included, and the damages that will be compensated will understate true social costs.

Comment: I would like to pursue a little bit the problem within the news media. My experience with the news media, has been that the news has been very good for the environment. Many times scientists are more concerned about the scientific outcome of their studies than they are about informing the public. How was the news media treated during the *Amoco* case? In relation to technical details, Dr. Sorensen mentioned that some media reported figures on the actual costs five times as high as those figures he estimated. Have you actually talked with journalists who have given the figure of two billion dollars and are they misquoting you?

Sorensen: Answering the last part of your question first, the two billion dollar figure represents claims against Amoco. This compares to our figure of $100 million dollars. Once you get a two billion dollar figure quoted in *Fortune* and several other places, it becomes a "fact" which is often misrepresented. The claims made in a law suit are not an evaluation of damage, but a lot of newspapers or news agencies like United Press, for example, will see a number for legal claims and then report that economic damages amount to two billion dollars.

Comment: I think that must be stopped and could be stopped if you were actually quoted in the news in a right way, because after all, you were the chief investigator. You can correct the figure, the same as we have in the Scandinavian countries.

Sorensen: Our preliminary numbers, the ones you heard today, were developed over the last eighteen months, but only presented for the first time at this conference. We couldn't very well have corrected *Fortune* magazine, which came out in April, 1979 with their two billion dollar figure. Yes, we will be trying to reach the press when we get final numbers. The numbers you heard today are very preliminary and nobody on the study team wants numbers going out that will have to be contradicted later, when we finish our study. So we have to wait a few more months for the final product which will include other categories of damage. I've only indicated early pieces of the puzzle here today.

Ultimately when we finish our work and when it is delivered to NOAA, we plan to have a major conference to do all that we can to get our results out, because I think they will have important policy significance.

Comment: Regarding the last question, I'm increasingly struck by the crucialness of the damage function that you have to establish in formulating your econometric mode.

I'm also struck by the role that this total study will have on the subsequent compensation which is to be awarded and how it might influence the awarding of damages. Why do you propose trying to insulate yourself instead of forcing some role in the ultimate dispensation?

Sorensen: What you say is true. The numbers that come out of an academic study of this kind are likely to be cited in a court case. This possibility has been discussed by the whole team of researchers. Our French research colleagues and certainly the government of France has recognized the possible problems. None of us likes the fact that we may have to appear in court or give depositions on this subject.

A very important point is that there are major differences between legally compensable injuries and social costs. Some series of businesses may come into court and recover large legal damages that are not social costs, because they are offset by transfers to other regions in France. In fact, we can show that other regions of France picked up recreation traffic that would otherwise have occurred in Brittany in 1978. That doesn't, however, stop the hotels and the restaurants of Brittany from coming in and claiming damages, assuming other procedural issues can be settled.

Comment: It's probably way too early to give a definite answer, but I'd like to know if Mr. Sebek or Mr. Laubier could tell us about the *Tanio* oil spill in relation to the coming into force of the IMCO compensation and liability fund in between the *Amoco Cadiz* and the *Tanio* spill. Has there been any difference in the social or legal impacts of those spills?

Sebek: I could say that the Assembly of the Fund examined *Tanio* at their meetings. The claim is still pending. One of the things which the fund specified is that settlement for the party depends on how reasonable were the costs incurred in clean up by the French government. The 1969 Civil Liability Convention also specifies that the cost must be reasonable.

I would also point out that the claim for the *Elena V* accident which occurred in 1978 off the British Coast is still pending because it is widely argued that the British government rather unwisely spent lots of money on dispersants which are ineffective on heavy fuel oil. This, incidentally, was the same kind of oil that was spilled from *Tanio*.

Laubier: I would like to add something to what you said from the social point of view for the *Tanio* spill, regarding psychological effect. At the time of the *Amoco Cadiz*, we had a lot of volunteers for cleaning work, including people from the area, from other parts of France, and even from foreign countries. Also, during the *Amoco Cadiz* oil spill the scientists were able to control which dispersants were to be used to clean some parts and prevent the use of detergents. Dispersants were used in normal quantities far from the coast.

The case of the *Tanio* was different, which may be explained by the fact that the people are already nervous and angry. There was no volunteer effort at all at the time. Nobody wanted to go by himself and spend his time cleaning as a volunteer. In addition, the French authorities have not been able to control the use of chemical dispersants. In a few cases, normal detergents you would put in a washing machine have been used in huge quantities to clean rocky areas in the Côtes du Nord.

Waldichuk: I've heard stories that the European eel develops tumors when exposed to oil. Did you find this happening at all? Was there any indication of tumors in the European eel?

Laubier: Yes, they have definitely been found. Also found in the European eel is an accumulation of oil within the gonads, together with a complete degeneration of the ovocytes. Also, unusually large activity within the gills and the inter-renal gland has been observed, which probably means that the animals were still in a stressed condition, one year after the wreck. The eels were from the Roscoff area.

UNIT THREE

"Intentional" Pollution: A Case Study of Ocean Dumping in Long Island Sound

Photo courtesy of Prentice K. Stout

THE TITLE OF THIS UNIT IS "INTENTIONAL POLLUTION: A CASE STUDY OF OCEAN Dumping in Long Island Sound." It is my firm belief that this title is misleading, and its use in a context such as this forum deliberately promotes adverse societal opinion. It implies that all ocean disposal of dredged material is intentional pollution. A title which connotes this is considered premature in a forum which is hopefully objective in its analysis.

An analogous title to "Intentional Pollution: A Case Study of Ocean Dumping," might be "Intentional Sex: A Case Study of the Original Sin." It is clear that the latter title is misleading, because it implies something bad by title alone, whether or not you have done some experiments with this, or whether or not you believe in it. The same is believed to be true in the first instance as well.

Just as I believe that modern society has evolved somewhat in its attitudes from the biblical times which depicted the activities of Adam and Eve, and their perception of sex guilt after eating fruit from the Tree of Knowledge, so I hope that the scientific community is gradually becoming free of preconceived prejudices, the latter from eating generously from the scientific tree of knowledge.

In any event, significant progress has been made in the last decade concerning dredged material disposal. This has generated an enormous literature on all aspects of the disposal of dredged material at sea. The overall conclusion from this is that the adverse environmental effects appear to be minimal.

Of particular importance in this respect is a program known as DMRP, Dredged Material Research Project, sponsored by the U.S. Army Corps of Engineers, Waterways Experiment Station, Vicksburg, Mississippi, a program that amounted, as I recall, to more than $30,000,000 of research money over a 5 year period.

The case study of ocean disposal of dredged material to be considered involves a portion of the DAMOS Program, a very sophisticated monitoring and applied research study in the New England area. This study is largely, at present, under the supervision of Dr. David Shonting, a physical oceanographer who is also a member of the faculty at the Graduate School of Oceanography, as well as being associated with the Naval Underwater Systems Center, Newport, Rhode Island.

Lance Stewart, a biological oceanographer from the University of Connecticut, is going to offer some biological aspects of this DAMOS Project, as he has been actively involved with it for some time.

In addition, Chris Roosevelt, who is president and director of the Oceanic Society and who has been actively involved as an environmentalist and as a concerned citizen dealing with aspects of both the DAMOS Project and subsets of it in the Long Island Sound, will discuss dredging and spoil disposal policy.

Even I have been involved to some extent with these studies in a modest advisory capacity.

Saul Saila
University of Rhode Island
United States

CHAPTER 9

The New England Disposal Area Monitoring System and the Stamford-New Haven Capping Experiment

DAVID SHONTING
Naval Underwater Systems Center
United States

ROBERT W. MORTON
Science Applications, Inc.
United States

INTRODUCTION

In the late thirties and early post war years, the increasing demand for navigable channel dredging plus tremendous growth in industrial waste production forced a widespread use of ocean "dumping grounds" or "dump sites." These were usually rectangular areas of 2–40 km² located a few kilometers off the coast, yet close enough to industrial waste producers or harbor dredging areas for economical transport. It was still widely believed that materials dumped in the "open" ocean were safely isolated from man's environment. Increased utilization of such areas was a first step in attempting to come to grips with the growing industrial waste and dredge spoil problem, but it was not the solution.

During the 1960s and 1970s there developed an exponential increase in awareness and concern by the public regarding waste disposal on land and in the sea. Indeed, within the past fifteen years new organizations as well as laws at federal, state, and municipal levels have been created to oversee and strongly regulate dumping of dredge spoils and industrial wastes. Industry, today under great pressure from these statutes, is searching for ways to eliminate, store, or dispose of its wastes.

The dredge spoil problem, however, is unique in that as long as there is maritime transportation there will be a need for dredging of navigable channels, and there appears no alternative but to return the spoils to the marine environment. Where this disposal is to be placed in order to produce minimal adverse environmental impact must be determined. One favorable factor is that the industrial waste disposal in tributaries and harbors is diminishing. This will reflect in progressively "cleaner" dredge material and should reduce the opposition to dredge disposal in the future. However, this aspect is likely to be many years away.

The principal concern of critics of marine disposal of dredge spoils is the possible degrading or toxic effects upon marine life on the bottom and within the water column. Moreover they ask, if there are immediate or long term detrimental influences upon the biota, to what extent are these effects then passed along the food chain to inflict damage upon sport and commercial fishing industries and, ultimately, on public health? It is generally agreed that the burden of proof that dredge disposal sites are or could be utilized safely must lie with those responsible for conducting and managing dredging operations.

We are thus faced with the necessity of disposing potentially toxic dredge materials, while at the same time we must gain an understanding of disposal site dynamics in terms of physical, chemical, and biological parameters in order to evaluate the potential for any adverse environmental effects. It must be emphasized that only with these data can we make the best possible decisions to minimize those risks while permitting the necessary dredging operations. The proper course of action is, then, to gain a clear scientific understanding of the problem; and to present these facts to the public along with appropriate plans of action to insure their protection while allowing the continuation of ocean dumping which is continuously guided by programs of applied research and monitoring.

It was with these problems in mind that the New England Disposal Area Monitoring System (DAMOS) was developed as a joint effort of the U.S. Army Corps of Engineers (COE) and the Naval Underwater Systems Center (NUSC). The Navy Laboratory became involved in dredge spoil studies through requirements generated by construction of 688 class submarines in New London, Connecticut. The additional draft of these vessels required substantial dredging and disposal of material from the Thames River. Therefore, the Naval Facilities Command requested that NUSC oceanographers coordinate studies of possible dump sites with the U.S. Army Corps of Engineers who deal with permits and dumping controls.

BACKGROUND AND OBJECTIVES

The objectives, direction, and scope of the DAMOS program were formulated in 1977 from the experience gained from two years of sampling and analysis of dredge material disposal sites off the New England coast. In reviewing this work, it became obvious that a more unified assessment of disposal sites and their environmental effects and a more efficient and cost effective approach to obtaining this assessment were needed.

Fundamental to DAMOS, the program is the creation of a "regional approach" to monitoring disposal sites. Prior to the initiation of the program, funding for dredge spoil disposal site monitoring was provided to several different laboratories to study separate sites. This work produced a series of reports varying in content, quality, and completeness, often depending on the size of the dredging project and the particular strengths and weaknesses of the funded institutions. Through the regional approach to monitoring and the consolidation of the technical management, the development of more sophisticated data acquisition systems and the standardization of measurement techniques was accomplished. Furthermore, all environmental data was consolidated to a common base. This approach continued to encourage the use of the expertise available in the individual laboratories in such a manner as to apply specific institutional strengths to regional investigations.

Monitoring of regional disposal operations must be based on EPA criteria set forth in sections 228.9–228.13 of the Ocean Dumping Act (1977) which contained guidelines for measurement of physical, chemical, and biological impacts of dredge material disposal in coastal waters. However, under the DAMOS program, these criteria were used primarily as guidelines in order to accommodate the economic realities of a regional program and to permit flexibility in response to specific problems associated with disposal sites in the New England area. In practice, the criteria are divided into two major monitoring studies: the first, a physical measurement program to define the geometry and evaluate the stability of spoils in the disposal site, and second, a biological-chemical program designed to discover and evaluate any adverse impacts on biota, both within the disposal area and in the coastal environment as a whole. Since the biological-chemical studies were to be strongly dependent upon results of physical observations, the DAMOS project provided a multi-disciplined program with stressing continuous interaction among scientists from various fields of study.

Dredge spoil disposal in the coastal environment is often a politically sensitive and controversial issue. Some criticisms of dredge spoil operations are well founded particularly where faulty decisions were made because of a lack of data and improper coordination of information. However, results of the Dredged Material Research Program, conducted by the Army Corps of Engineers, have shown that disposal can be accomplished with minimal damage to the environment if informed, prudent decisions based upon solid data are made relative to the disposal operations. In order to meet these critical data requirements, management of disposal operations must be supported by meaningful, consistent data acquired with the best available instrumentation systems. These data, the associated systems, and measurement techniques must always be available for scrutiny and criticism by government agencies, the scientific community, concerned environmental groups, and the public at large. Consequently, a major emphasis in the DAMOS program has been presentation and open discussion of the approaches, measurements, and analyses used to monitor environmental impact. Dissemination of information has been accomplished through reports, an annual symposium, public hearings, meetings with government agencies, and presentations at professional conferences.

The DAMOS program continues to be of extreme importance to the Corps of Engineers (COE) because of the following. It represents a continuous effort to insure that impacts resulting from dredge spoil disposal are small, provides data for assessment of disposal permit applications, aids in the management and operation of both large and small dredging projects, and supplies factual information to concerned environmental groups to satisfy regulatory criteria imposed by other federal agencies. Moreover, the program has increased the public's awareness of the effort made by the COE to better serve the interests of environmental protection while at the same time carrying its objectives to administer public waterways.

The DAMOS program has also made important contributions to research programs at the Naval Underwater Systems Center by the development of new measurement techniques and instrumentation systems to study shallow water environments which have significant impact on fleet and weapons system exercise procedures. In particular, navigation and data acquisition techniques developed by DAMOS are now incorporated in many Navy test and evaluation programs and in underwater search missions.

THE DREDGE DISPOSAL SITE AS A DYNAMIC SYSTEM

A principal aim of the DAMOS program is to gain understanding of how a disposal site containing "foreign" material interacts with its surroundings. In order to define the approaches to the study of a spoil site, it is helpful to consider it as a physical and geological entity within the ocean environment. A typical modern dredge disposal area in the New England region (usually labeled "dumping ground" or "dump site" on navigation charts) is a rectangle measuring 1–10 kilometers on a side and having water depths ranging from 20–50 meters. Until recently the delineation of dumping grounds by marks or buoys was inconsistent, sometimes two or more buoys were used and in other areas none. This resulted in dump materials being spread over areas with a minimal control of the site location. In recent years, as a result of DAMOS operations, dredged material is more often released by use of modern navigation systems in close proximity to specially constructed dump site buoys.

Assuming that dumping is conducted with consistent navigation fixing at the marker buoy, a typical disposal site (fig. 1) may be a circular or eliptical shaped mound of material 300–800 meters across and may reach 3–8 meters above the mean local depth. The volume of the mound may range from 50,000 to 500,000 m³ (500,000 m³ is roughly the volume of a right parallelepiped with a football field as a base and a height of 400 feet). Probably the single most important characteristic of a spoil site is its ability to hold or contain deposited material, i.e., its degree of "containment", other related terms being its stability or erodibility. It is upon the parameters that affect the sites containment that we concentrate our study.

Dredge disposal sites are usually located in relatively shallow coastal waters. The oceanographic implications are that the overlaying water column is very dynamic and that (as suggested by fig. 1) perturbations on the spoil mound are imposed by strong tides and meteoro-

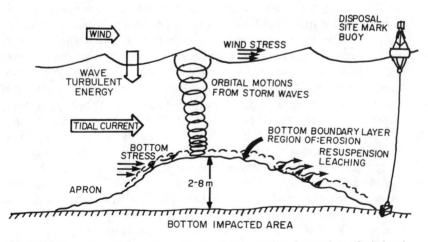

Fig. 9.1 Schematic portrayal of a spoil site, showing various interactions of wind and tide that produce motions that can erode the surface of the spoil area. Marker buoy is taut moored to provide reference to dumping barge.

logical events. Since the amount by which the spoil mound is disturbed or eroded governs its interaction with the local environment, the effects of these geophysical phenomena must be measured.

Another important factor governing the degree of containment and the "erodibility" of a dredge material is its physical-geological character at the time it is deposited from a dumping barge and after the physical changes which it may endure after deposition on the sea floor. The physical character of dredge material is determined by its composition, its grain size distribution, degree of compaction, porosity, and water content. Furthermore, spoils contain varying amounts of organic material that can significantly affect cohesive properties.

The greatest potential for interaction of dredge material with the water column occurs at the time of release as the sediment cascades downward through the water column. However, studies have shown that the material falls as a density current and a large percentage of the material impacts the bottom within one to two minutes after disposal. A small plume of fine material is often generated which disperses and is advected from the disposal site in several minutes.

Once the spoil material impacts the bottom, the ambient water motions may affect the deposit to an extent depending upon the character of the resident material, how it is physically distributed on the bottom, and on the motions of the water. Erosion of the bottom material can only occur if the kinetic energy and Reynolds stresses of the motions at the bottom can overcome the threshold level of cohesive forces and inertia of the material.

Turbulent or oscillatory motions sufficient to cause erosion may be associated with strong tidal currents which produce a shear zone (boundary layer) near the bottom and with large surface waves whose orbital motions extend clear to the sea bottom. Abnormally strong tidal currents and large surface waves are associated with storms and hurricanes and it appears that most of the significant sediment motion on New England disposal sites may be caused by these spurious events. When the bottom material is brought into suspension, it is then moved horizontally by the ambient currents. The material will begin to resettle to the bottom when the turbulent mixing forces become less than those of gravity. Since the energy required to erode material is significantly greater than that required for transport, any material eroded is carried well beyond the disposal site, dispersed, and mixed with natural suspended sediment before it is redeposited. Consequently, the possibility of tracing such a material is nil since it is rapidly lost in the background levels.

During the period of active dumping, the topography of the site is continuously changing, possibly altering its surfacial character. Further, new dredged material may be added from different source areas, thus the site may contain several strata of varied composition and cohesive character.

Another effect on the spoil pile which can promote erosion are biological processes caused by the macrobenthic animals which burrow in and out of the spoil material. This "bioturbation" changes the pile surface, in some cases reducing the size of microtopography and homoginizing the upper layers.

The portrayal of the spoil site as a dynamical system emphasizes the need for hydrodynamical studies. Chemical and biological programs which traditionally dominate all pollution studies can only register the net results of material transfer, whereas prediction of the occurrence and magnitude of such material transfer into the environment lies in the understanding of the dynamics of the spoil site.

STRUCTURE OF THE DAMOS PROGRAM: FIELD OBSERVATIONS AND STUDIES

In planning the field measurement program, it was necessary to define a consistent set of parameters to allow proper intercomparison of the various disposal sites. The original sites targeted for the DAMOS program (fig. 2) were distributed along the New England coast from Long Island Sound to Northern Maine. Some are "active sites" (white circles) presently under use, and some are "inactive" (black circles) not having had material deposited in two to five years. Inactive sites may become active in the future, if dredging is required and new permits are obtained.

The character of coastal dump sites varies depending on their geographic locations, depths, and oceanographic conditions. Sites in the Gulf of Maine and Rhode Island Sound are facing the open ocean and hence experience heavy sea conditions and large waves especially during storms. Here the bottom environments may suffer erosive effects due to turbulence and currents associated with surface windwaves and swell. The Long Island Sound sites experience less effects from windwaves because of short fetches and the total absence of swells, but may be more affected by the tidal currents which tend to be stronger than occur on the open coasts.

One important property of a disposal site which is ascertained from baseline data is its degree of "containment," i.e., the extent that dredge spoil material is held or contained in some sites relative to others. This is partly due to the type of spoil deposited, i.e., its density and compaction. Moreover, the degree of containment will be a function of depth, gross topography, intensity of tidal currents, or surface wave action perturbing the spoil site. Clearly, knowledge of parameters fa-

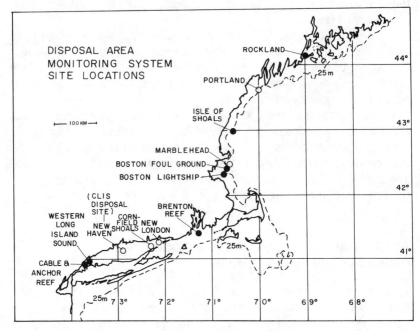

Fig. 9.2 New England disposal sites. Inactive sites (not used within two years) are dark circles.

voring containment characteristics would be useful for choosing future sites where high containment was desired.

One objective is to measure interactions of the spoil site with its environment. Essentially, the interactions or effects are deduced from observed "changes" in time (and space) of the spoil site and its surroundings. Hence, we must establish a "baseline" from which to refer our data as it is gathered. At the outset of DAMOS study there were no comparison data available by which to define changes, so it was necessary to create a bank of information describing the physical, chemical, and biological characteristics of the disposal site. These parameters include: topography, roughness characteristics, depth, approximate volume, material type, benthic community structure, and visual records of the site. To this baseline data were also added observed ambient oceanographic conditions which include: tidal currents and range of height, seasonal ranges of expected wind, and wave conditions, the degree of sheltering from the open sea, and topography of the general area.

The schematic structure of the DAMOS program is shown in fig. 3 with the program being divided into four parts:

SCHEMATIC PRESENTATION OF MONITORING PROGRAM

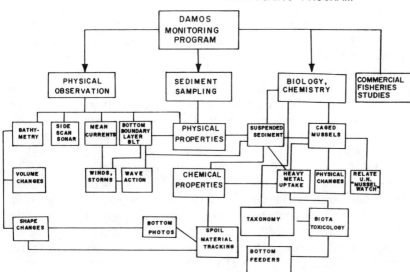

Fig. 9.3 Structure and present activities of the DAMOS Monitoring Program. Types of biology and chemistry studies may vary in time, depending upon changing emphasis in analyses desired.

Physical Observations: This includes measurements of the physical topography and character of the smaller scale features of the spoil site and its surroundings and observations of the dynamic motions of the currents and waves which affect the disposal area (45% of the project effort).

Sediment Sampling: Collecting and photographing samples of spoil and surrounding sediments is done to study the physical-geological properties as they experience effects of erosion, settling, and re-distribution. Additional chemical samples are obtained to charac-terize the dredge material and estimate its potential for affecting benthic biota (15% of project effort).

Biology and Chemistry: Collections and analyses are made of spoil and surrounding sediment samples for heavy metals, toxicity, and effects upon biota living within the sediments. Further, collection of various species of bottom fauna are made to assess populations in and around the spoil sites. Mussels placed in the spoil sites are tested for uptake of heavy metals (35% of project effort).

Commercial Fisheries Data: Study is made of the commercial fishing existing in the vicinity of the disposal sites. Study is made of variety and size of catch as a function of time of year and at various locations. This data can provide background information in ascertaining effects of dredge spoils upon commercial fishing in a particular area (5% of project effort).

Communications of the results of the DAMOS studies are produced in technical reports (see Appendix), and publications by the various principal investigators in referred journals. DAMOS data files are held by laboratories of Science Applications, Inc., Newport, RI, the Graduate School of Oceanography of the University of Rhode Island, Kingston, RI, and the Marine Institute of the University of Connecticut, Avery Point, CT. Presentations are also made at numerous public hearings sponsored by the U.S. Army Corps of Engineers or state and federal agencies.

Finally, as part of the Commercial Fisheries Study (Pratt 1979), communication is made with fishing companies, co-operatives, and with individuals to ascertain possible effects of spoil disposal upon the fishing industry. Also, dumping schedules are coordinated with fishing activities to have a minimum interference with seasonal cycles of migration, spawning and hatching of both shell and fin fishes. Although the fisheries study is a small part of the DAMOS program, its importance to its success as a means of public relations cannot be over emphasized. Communications and mutual interchange of data and ideas between DAMOS workers and fisheries and economic interests is vital both to guide the direction of the studies and to accelerate application of the knowledge to practical utilization by both the users and the protectors of the marine coastal environment.

THE DISPOSAL SITE

Physical Methods of Observations and Analyses

The following is a summary of techniques instrument systems developed to conduct both the DAMOS baseline surveys and monitoring programs:

Navigation Control System

It was necessary to obtain precise baseline bathymetric data of the spoil sites and further, if possible, to monitor topographic changes as barge dumping proceeded. This imposed stiff requirements both for depth resolution and horizontal positioning, i.e., navigation control.

A bathymetric data acquisition system (BDAS) was developed (fig. 4) to provide the required navigational control (Massey 1979). A general

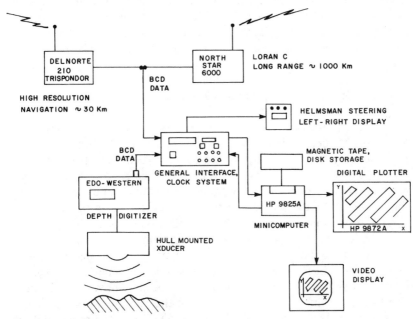

Fig. 9.4   Schematic of the BDAS navigation system used aboard the NUSC survey boat R/V *East Passage.*

interface system utilizing a precision clock transfers precise position data from a high resolution Del Norte Trisponder navigation system into a Hewlett Packard minicomputer (HP 9825A). The computer provides real time magnetic tape recording, a video display, and a digital plotting of the ship's track along with updated information on range in meters and bearing to a reference buoy, course made good, and time of day. The resolution of the system to detect relative changes in position is within 1–3 meters depending upon the distances from the Del Norte shore based transmitter-slave stations (nominally 20–50 kilometers).

For conducting fine scale grid-based surveys or searches the BDAS encorporates a Helmsman's Aid display. This is mounted on the bridge of the survey vessel to provide the steerer continuous guidance by port/starboard lights indicators to maintain a prescribed true course over the bottom.

The BDAS can also utilize Loran-C navigation signals using a Northstar 6000 system. This provides navigation to within 10–50 meters over long ranges from 50–500 kilometers from shore, eliminating the need for establishing shore stations. The BDAS has been modified by Scientific Applications, Inc. to eliminate the interface system by use of an Apple II minicomputer. It has increased portability and it can be used in the field by personnel with no special computer training. (A system was used successfully for several weeks by tow boat personnel to make routine dumps at a precise location at the New Haven disposal site.)

Bathymetric Data: Acquisition and Presentation

The BDAS (discussed in the previous section) couples an Edo-Western and Digitrac (depth digitizer) with the precision navigation and recording capability (fig. 4). Binary coded decimal (bcd) data from both the precision navigation and the echosounder is matched to the precision clock and fed to the minicomputer and hence to the tape recorder while the plotter and video display portray the ship's track in real time. The computer is programmed with the desired survey tracks (fig. 5).

The survey vessels actual track plotted in real time is shown superimposed on the grid (fig. 6). The Helmsman's Aid places the ship generally within 2–3 meters of the survey lanes depending upon the operator's skill and the sea conditions. As the ship moves, the digitized depths are obtained at the update rate of the fathometer and averaged over the update rate of the Trisponder (1/s). For a boat speed of 3 m/s (6 knots) this provides bathymetry data points at 3 meters spacing. Note that the depth data are recorded at ships position irrespective of the lane position, i.e., the lane map only serves as a guide to provide optimum data distribution for contouring and other data presentation. Prior to chart preparation, corrections are programmed into the computer for transducer depth, mean speed of sound for the water column, and the state of the local tide.

For the most precise portrayal of changes in topography of the spoil pile (which may be caused by addition of material, erosion, redistribution, or settling) is by obtaining running identically positioned lane profiles. By superposition of sequential repetitive profiles changes in depth can be resolved down to 10–20 centimeters.

The recording of the original data on digital magnetic tape allows utilization of a variety of innovative data display and graphics available with modern minicomputer systems. Bathymetric data of spoil sites are routinely computer plotted as fine scale contour charts with 0.2–0.5 meter contour intervals and standard 75–200 m/inch scales for intercomparison with previous or future charts. An example of New London site is shown in figure 7. In addition, isometric bathymetry maps are computer generated which present a three dimensional picture of the spoil site. Here is shown the New London site looking 45° downward across the lower right hand corner of the site (fig. 8).

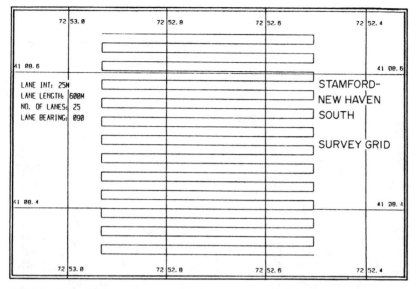

Fig. 9.5 Example of actual survey grid programmed into HP 9825 computer and machine plotted for planning reference.

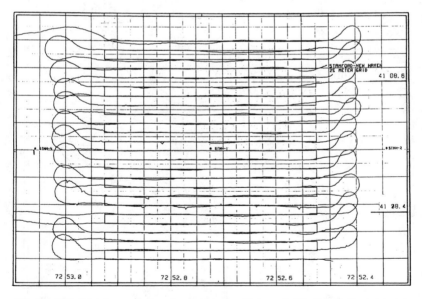

Fig. 9.6 Ship's track superimposed upon survey grid shown in Fig. 9.5. Bathymetric data is provided each second along ship's track as it moves over grid, guided by Helmsman's Aid.

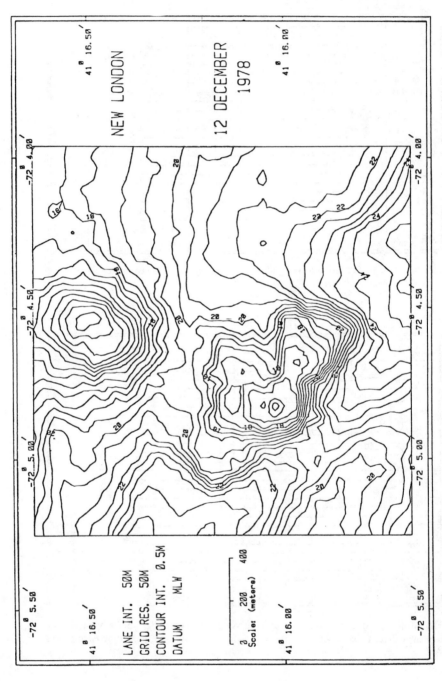

Fig. 9.7 Example of computer-drawn contour map of New London spoil site.

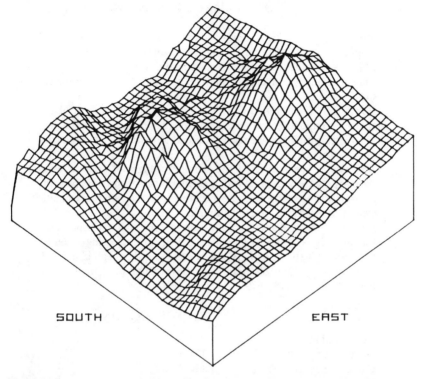

NEW LONDON
6-8 MARCH 78 SURVEY
VERTICAL EXAGGERATION - 60
VIEW ANGLE - 45 DEG

SOUTH   EAST

Fig. 9.8 Isometric portrayal of New London spoil site bathymetry shown in Fig. 9.7. Drawn with HP 9825 computer, driving HP 98 9872 machine plotter.

Calculation of change in spoil volumes can be done by computer integration of the differences between successive contour charts. The error in calculation is associated with the resolution of the bathymetry which for 25 meters land spacing over 700 × 700 meter grid is on the order of 2000 m³ (i.e., 4% of a 50,000 m³ volume).

Conventional Current Observations

Long term current measurements have been made at the sites to characterize the influences of the tides and high wind conditions upon the currents (Ianniello 1979). ENDECO-174 current meters, which internally record speed and direction, are tethered by a 2 meter line to a fixed mooring and are dynamically decoupled from vertical mooring motion rendering them more suitable than Savonius rotor systems for near surface observations.

Records were obtained at mid depth for periods of at least one month to monitor the basic tidal character, and to register any high energy storm events. The following data analyses were performed:

Motion Ellipses. This technique allows evaluation of the tidal and nontidal, residual motions by means of forming the tidal ellipses which depict the variation of the current strength and direction while providing convenient intercomparison of data at different sites.

Horizontal Kinetic Energy. This is estimated from squares of the velocity components using peak and average maximum tidal velocities. Sites with lower values would tend to be more favorable containment areas.

Anomalous Events. Periods of extreme changes of amplitudes in the currents are correlated with high energy storm and wind periods. Information on such events are available from the National Climatic Center in Asheville, North Carolina.

Ranking of Sites. A relative indication of the containment potential for all sites is obtained by ranking them in decreasing order of horizontal kinetic energy. Below is presented the mean kinetic energy from the total observed time series from sites identified in figure 2 during 1978–1979:

Site	Kinetic Energy (erg/cm^3)
New London	333.0
Western Long Island Sound	184.3
Cable and Anchor Reef	134.7
New Haven	125.7
Portland	12.5

The New London site lies near the "RACE," a channel known for its strong currents often exceeding 150–250 m/s. The remaining Long Island sites, although more sluggish than New London, are still at least an order of magnitude higher than Portland, which lies in open water. By use of the motion ellipse calculations, we can estimate the fraction of the total motion associated with tidal energy for each site as given below:

Western Long Island Sound	0.90
New Haven	0.86
Cable and Anchor Reef	0.78
New London	0.78
Portland	0.52

Further studies of the tidal and nontidal energies and their relationship with the boundary turbulence will provide insight into the effects of wind and storm conditions upon the currents and erosive and containment properties at each site.

Bottom Boundary Layer Turbulence Studies

The currents and wave action are known to be the significant parameters affecting the erosion of the surface material of the spoil mound. However, theoretical predictions of erodibility of bottom material are very unreliable because of a lack of data from which to construct models. Furthermore, observational evidence is mostly derived from laboratory tank tests in which it is extremely difficult to imitate ocean conditions.

Precise measurements are required of the motions near the sediment-water interface where the fluid flow is altered by the spoil mound, i.e., the bottom boundary layer. Supplementary to the bottom measurements, we must utilize the data of the tidal and mean currents (discussed above) which should be related to the turbulent energy at the bottom and also serve to transport spoil material which has been resuspended.

For the boundary layer turbulence studies a special instrument package (BOLT) was developed to measure small scale velocity fluctuations in the bottom boundary layer at various sites during both normal and storm conditions (Ianniello 1979). From these data, instantaneous shear Reynolds stresses and kinetic energies can be estimated. These parameters, when compared with critical erosion data from both in-situ and laboratory tests, will help to predict the potential dredge material erosion. The ultimate goal of the BOLT program is to characterize the bottom stress environment in terms of readily measured or predictable flow variables such as tidal current, wave regime, depth, and bottom topography.

The BOLT system utilized small ducted impeller meters designed by Smith (1978) and applied by Shonting and Temple (1979) for measurements of surface wave orbital motions. The impellers respond as the cosine subtended by the rotor axis and flow direction. The impeller systems provide a voltage output related to the angular velocity of the impeller. The signals pass through a microprocessor system (Shonting, Roklin, and Temple 1979) where it is digitized and stored on a cassette tape system which upon recovery is transcribed on to a standard nine track tape for analysis.

The BOLT system (fig. 9) contains arrays of sensors in fixed orientations off the bottom by a vertical rod which is suspended from a rigid tripod (fig. 9 insert). The system is lowered from a ship and oriented at the bottom by divers. The orientation of the impeller systems (fig. 9) provides a measure of the three fluctuating velocity components n',

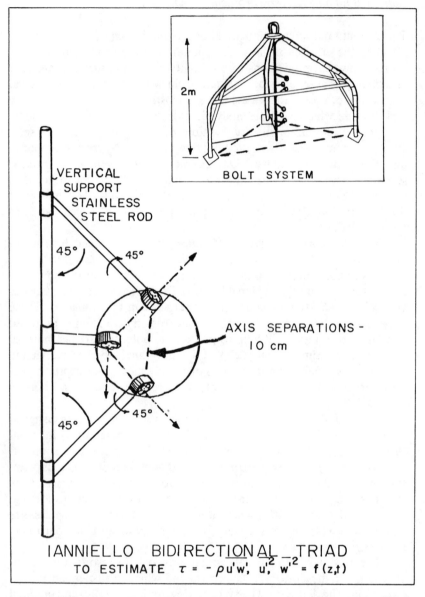

Fig. 9.9 BOLT System. Tripod mount for bottom suspension of sensors (insert) and array of three mini-impeller meters arranged to sense three-dimensional motion in the bottom boundary layer.

r', and w' in the x, y, and z directions respectively. These data provide estimates of the Reynolds stress $T = -\rho \overline{u'w'}$ where ρ is water density and the overbar represents a time average (which may be varied according to the particular time scales to be examined).

Figure 10 shows preliminary results of both Reynolds stresses estimated on a continuous basis for a 560 s interval. The positive value indicates that horizontal momentum flux is downward which provides the energy to erode the spoil material. The middle curve shows the horizontal fluctuations u' superimposed upon the mean $\overline{u}$. The lower curve depicts the vertical fluctuation w'.

These BOLT observations will be correlated with actual bottom

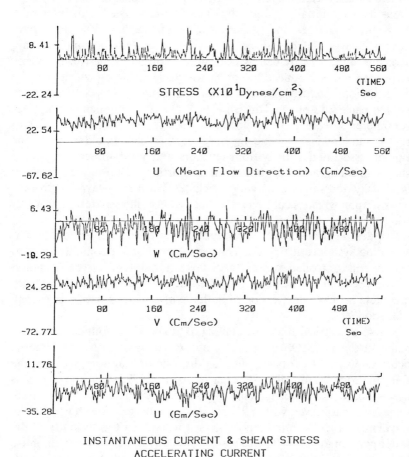

INSTANTANEOUS CURRENT & SHEAR STRESS
ACCELERATING CURRENT

Fig. 9.10 Sample output of Reynolds stress and velocity components obtained from mini-impellers.

erosion measurements being conducted at sites in Long Island Sound (Bohlen 1979). These observations are being made with a newly developed instrument array which periodically samples the water column immediately above the spoils interface for the concentration of suspended water while at the same time photographing the bottom and obtaining a measure of its transmission coefficient.

Sidescan and Vertical Sonar, Grab Sampling, and Scuba Diver-Television and Photography

The sidescan sonar offers a new approach to portraying the sea bottom by acoustically scanning and imaging in three dimensions relatively large areas as if by aerial photography. The sidescan system (manufactured by Klein Associates, Inc.) contains a high frequency (100 kHz) transmitter-receiver which is towed 5–10 meters above the bottom and maps acoustic reflections by beams transmitted normal to the tow direction. An echogram is produced by sophisticated signal gating which provides a continuous picture of the sea bottom from a 100–200 meter wide swatch each side of the Towfish. The 100 kHz frequency having a 1.5 centimeter wavelength can portray a variety of bottom types such as rocks, mud and sand, and even delineates small boulders and sand ripples associated with bottom currents and wave effects. Figure 11 shows an echogram of the bottom at the Portland disposal site; displayed is a rocky and sandy zone which includes 3–4 meter wave length sand waves, upon which is cut by a lower flat region of finer material, probably silt.

Sidescan surveys can be conducted as with the conventional echo-sounding system utilizing the Helmsman's aid and a computer-guided survey grid controlled by the BDAS navigation system. The combination of the precision bathymetry plus the side scan records provide a very complete acoustic description of both the maco- and micro- topography of a spoil site.

Use of vertical high frequency (200 kHz) echo sounder has been made in the DAMOS program to observe dynamics of spoil material being deposited by a barge. During disposal of Stamford spoils at the Central Long Island disposal area in April, 1979, such a system was used aboard *East Passage* (the NUSC survey boat) to observe spoil release as it followed within 25 meters of the barge. The short acoustic wavelength (0.75 centimeters) allows it to portray the turbidity in the water column in detail. The releases were done at slack tide to minimize horizontal advection while observing the vertical transit of the material. The echograms clearly displayed the high acoustic density cloud of spoil

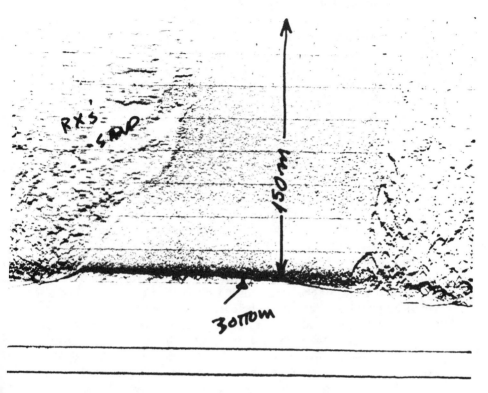

Fig. 9.11 Record of Klein sidescan sonar system from Gulf of Maine.

material cascading downward. In several dumping operations, it was indicated that most of the spoils material settled out of the water column within 1–2 minutes. This technique, however, has limitations since quantitative measurements of concentration of material left in suspension cannot be made acoustically. However, this method can be used to infer the pattern of deposition during disposal, especially when used in conjunction with the Sidescan Sonar to make immediate examination of the newly fallen spoil material.

The basic sampling tool to obtain sediment material from the upper 10–20 centimeters of bottom sediment or dredge material is the Smith-MacIntire Grab. The device provides rapid collection of samples weighing several kilograms, which are placed in polyethylene bags, and for biological sampling, can be refrigerated until analysis is made. Samples can be analyzed for physical properties such as grain size, density, and porosity, and for chemical content of heavy metals, and toxic inorganic materials. Furthermore, grab samples can help in analysis and interpretation of Sidescan Sonar records in which acoustical reflectivity characterized by the echograms can be correlated with physical attributes.

An extensive scuba diving survey program has evolved from the work of Stewart (1980) which demonstrates the need for visual and, wherever possible, diver observations of the spoil mound to correlate and better understand remote measurements and address questions that can only be answered by in-situ observation. The overall objective of this program is to obtain a chronological, photographic record of spoil surface conditions and benthic populations on and adjacent to active disposal sites before, during, and after disposal operations.

The following important activities were executed by a team of scuba divers:

a. Monitoring of benthic macrofauna behavior as recolonization of the spoil mounds occur which affects the spoil mound sediment character and overall stability of the spoil material.

b. Close examination of the spoil margin characteristics to provide more precise estimates of spoil bottom distribution and volume estimates, both by identifying dominant organisms and sediment characteristics.

c. Assisting in studies of spoil stability by deployment and photography of erosion stakes, diver navigation transect lines, and bottom sampling. Divers often examine and sample an area surveyed with a Smith-MacIntire grab to compare the results and estimate how representative a random grab sample is to define the bottom type. This is critical since in deeper areas grab samples are the only convenient means of bottom sampling.

d. Using hand-held underwater television to survey shallow sites where a video monitor and magnetic tape recorder record the data aboard ship.

Biological Sampling and Chemistry

The DAMOS program emphasizes the physical measurements and mechanisms of material transfer from the spoil site to the nearby environment. A comparably extensive biology or ecological impact study was not possible due to limited funding and organizational constraints. Clearly, however, certain biological and chemical studies were required to relate the dynamics of the disposal sites to observable contaminating effects upon the surrounding bottom sediments and bottom biota; the areas of study are indicated in figure 3.

Two studies undertaken were a survey of benthic microfauna in the disposal site environment and an assessment of heavy metal uptake by blue mussels (mytilus edulis) placed at or near the sites. These programs depend on the heavy metal analysis performed both on biota and sediments. A brief review of these programs follows.

Benthic Ecology

At the outset of the DAMOS program a benthic ecology study was initiated to provide as a baseline of seasonal data the population density and spatial variability of the benthic macrofauna population (Brooks 1979). Bottom samples are obtained with Van Veen grabs as well as dredge tows wherein the specimens are identified and counted along with an examination of the sediment type.

As a result of this sampling program, a relatively good description is available of the population dynamics and of the disposal site areas. The next step is to assess the effects of disposal on the benthic population (beyond the obvious catastrophic events of local burial). This requires post-dredging monitoring of benthic fauna to observe any significant deviations from the baseline data.

The basic problem in trying to assess spoil effects on biota is due to the inherent high heterogenous distribution or patchiness of fauna on the bottom. A strong local correlation, however, has been found to occur with the number and forms and the types of sediment which in certain areas (e.g., around Portland site) is strongly heterogenous in its spatial distribution. This high variability can produce a statistical noise which can mask effects of population density distortion produced by pollution effects. Discriminant analysis techniques based on substrata characteristics (Walker, Saila, and Anderson 1979) appears suitable to reduce variability among samples. Furthermore, cooperative and comparative studies will be made with the "Ocean Pulse" program of the National Marine Fisheries Service (Pearce 1980).

The health of the benthic macrofauna community is an extremely important criteria for judging the environmental impact of a dredge spoil disposal site and as such is closely involved with all other aspects of the DAMOS program. Thus, benthic data is compared with visual observations of the macro and megafauna by divers to obtain an overall understanding of community structures. Bathymetric and chemical data are significant inputs toward understanding the substrata of the benthic communities and therefore must be closely correlated with the macrofauna samples. Furthermore, the results of the current meter studies, suspended sediment data, mussel watch, and infauna programs all provide information on potential impact of spoils on the benthic macrofauna and must be considered in data interpretation.

Mussel Watch Program

Mussels and other selected shellfish are known to concentrate environmental pollutants in their hard and soft tissues and are thus monitors of water contamination and of the possible biological impact of dredge spoil disposal (Feng 1979). Although overt mass mortalities are not anticipated aside from the actual local burying of organisms during dumping, significant changes may occur during post-disposal at the tissue and cell level. For example, histological examination may reveal functional alteration of gonadal development, significant in terms of the perpetuation of the species.

At sites studied, mussels are placed adjacent to the spoil mound and at reference stations removed from the influence of the disposal operation. At most sites, the blue mussel (*Mytilus edulis*) is used except for deeper stations near Portland where the horse mussel (*Modiolus*) is utilized. Mussels are suspended in plastic mesh bags from cages made of FVC tubing. The mussels have been sampled on a monthly or bimonthly basis and analyzed for ten heavy metals: Cd, Co, Cr, Cu, Fe, Hg, Ni, Pb, Zn, and recent inclusions in the analyses will be polychlorinated biphenyls (PCBs).

At the New London site, where continuous sampling has been possible over a two year period, mussels have proven to be a sensitive environmental monitor for both heightened and lessened dredging activities as well as natural environmental events, e.g., river run-off. Therefore, with proper sampling and analysis they are potentially capable of monitoring any significant impacts on water quality caused by the presence of spoils at the disposal site.

As a result of the preliminary Mussel Watch studies, the objectives of the program will deal with the following questions:

> a. How do heavy metal levels in mussels near disposal sites reflect contamination from spoil material?
> b. What is the time response of mussel uptake and purging of heavy metals relative to initiation and cessation of disposal activity?
> c. What about "long term" effects of heavy metals upon reproductive and gill tissues in terms of injury and recovery?

The Mussel Watch program is closely related to the sediment chemistry and the suspended sediment studies since the mussels being filter feeders can take up contaminants from the spoil pile which have been resuspended by the turbulent excitation of the spoil material by the bottom boundary layer.

Sediment Chemistry

The sediment chemistry program evolved from an initial need to identify possible pollutants (namely, heavy metals) in the sediment and biota emanating from disposal site material. Initially, samples of mud worms (*Nepthys sp.*) and molluscs (*Astarte sp.*) were analyzed for Hg, Cd, Cu, Cr, Pb, Zn, Fe, and Co. A baseline of data was obtained from sites from the Gulf of Maine to Long Island Sound. Further use of the heavy metal analyses has been made to discriminate and characterize spoil materials in the benthic ecology studies and to provide the basic data to assess the potential impact of the spoils on the mussels and other infauna.

Recent studies at the Central Long Island Sound disposal area have shown that bulk heavy metals analysis of sediment samples can be a valuable tool for evaluating the result of a disposal operation (Jones 1979). Bathymetric data provides information on the geometry and amount of spoil deposit; however, precise horizontal extent of the spoil material around the mound can be delineated with chemical samples because of the significant difference between the amount of heavy metals in the spoils and background sediment. Samples obtained from spoil thickness of less than 10 centimeters can map the distribution of material and this can be correlated with visual and photographic observations and also effects of tidal currents and boundary layer measurements. The utility of this method was demonstrated on the Stamford-New Haven capping project (discussed in the following section) where spoils from each harbor were easily distinguished from background sediment and from each other.

THE STAMFORD-NEW HAVEN CAPPING EXPERIMENT:
A NEW APPROACH

During the past several years, attempts have been made to isolate small amounts of contaminated dredge material by covering it with cleaner material through specific techniques such as dredging from the head to the mouth of an estuary, dredging near docks and other "hot" spots during the initial phases of the operation, and combining it with clean material from nearby dredging areas at a common disposal point. These procedures have lead to the concept of "capping" in which contaminated materials are dumped at a specific point and then covered with a lesser amount of cleaner material.

One such capping operation was recently conducted by the New England Division of the U.S. Army Corps of Engineers (Morton 1980), at the Central Long Island Sound Disposal Site (fig. 2). A summary of this experiment follows.

The extreme shoaling conditions existing in Stamford and New Haven harbors required that dredging of these areas be conducted during 1979 to insure passage of commercial and particularly oil-related traffic to terminals in those cities. Bulk sediment analyses indicated that dredge material originating from Stamford harbor would be rich in heavy metals. It was decided to attempt to "cap" this material with silt and sand obtained from New Haven harbor in order to isolate the contaminated material from the benthic fauna and the overlying water column. Evaluation would be made of the relative merits of sand versus silt as capping materials in terms of coverage, stability, effectiveness in isolating contaminants, and recolonization potential of bottom fauna.

Monitoring of this disposal operation was conducted as part of the DAMOS program and consisted of the usual precision bathymetric mapping of soil distribution, visual observations of the spoil surface and margins, chemical comparisons of spoil and natural sediment and sampling of benthic populations for recolonization and bioaccumulation studies.

In order to compare the sand and silt caps, two disposal points of Stamford spoils were designated about 800 meters north and south of the spoil mound created by the New Haven project in 1974 (fig. 12). The south (S) site was to be capped with silt from the New Haven inner harbor and the north (N) site with sand from the outer breakwater area. The north-south orientation was selected since tidal flow through the

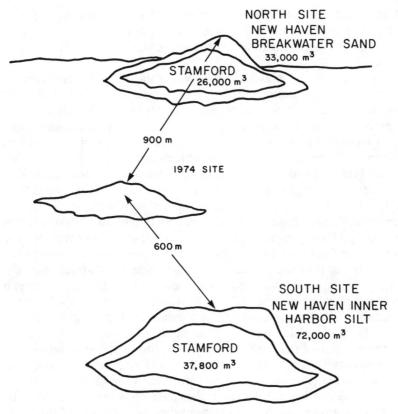

Fig. 9.12 Location of test spoil sites for capping operations.

site is in an east-west direction, thus potential contamination resulting from any erosion from the older mound would be minimized.

At the completion of a baseline bathymetric survey of each area, disposal commenced. Precisely located marker buoys were located for dumping references at each site. Initial disposal of Stamford material took place between March 25 and April 22, 1979 at the S disposal point. After April 24, silt from New Haven was dumped at the S site to provide capping material. Stamford spoils were then deposited at the N site. Disposal of silt at the S site and Stamford spoils depositing at N site continued until June 15 when silt and spoil dredging was halted to prevent damage to local oyster larvae by siltation. From June 15 to 21 a hopper dredge removed sandy sediment from the north of the New Haven harbor and used this material to cap the N site.

Disposal of dredge spoil from Stamford harbor at the S site reached a total of 37,800 m³ (based on scow load records) on April 22, 1979.

A survey of the S site on April 24 to determine the distribution of spoil material prior to capping indicated that the disposal had developed a small mound approximately 100 meters in diameter and 1.25 meters thick.

Calculations of total Stamford spoil at the S site detected relative to the baseline survey accounted for approximately 90% of the estimated volume deposited. Contour difference charts, however, indicated that there was appreciable additional material present beyond the immediate spoil mound, and it was possible that significant amounts of spoil had not been detected by acoustic measurements. This problem was addressed through a combination of visual diver observations and precision (50 meters spacing) remote sampling of the fringes of the mound with a Smith-MacIntyre grab. An extensive population of the stalk hydroid, *Corymorpha,* was found in the general area of the site. However, when dredge spoil was present to any significant degree, these hydroids were covered or destroyed; hence, divers could identify the boundary of the spoils by the presence or absence of these animals. Furthermore, the dark, organic spoils provided a sharp contrast to the natural, brown oxidized muds of the disposal site; thus, the thickness of spoils on the margins of the mound could be directly measured in the grab sampler.

The most striking result was the rapid decrease in spoil thickness at the margins of the spoil mound. In the east and west directions the change from thickness greater than 50 centimeters to less than 5 centimeters occurred between 100 and 150 meters from the disposal point; while in the north-south direction, the change was between 50 and 100 meters. In either case, it was apparent that the cohesive nature of the spoil material was creating a definite mound with discernable boundaries that could be detected acoustically to a spatial accuracy certainly better than 50 meters.

From these data, it is apparent that for cohesive spoils dredged with a clamshell bucket and transported by scow, most of the sediment (80%) falls to the bottom as a cohesive unit and forms a mound. The remaining material forms a turbidite type deposit radiating from the disposal point and the coarseness of the particles in the fringe areas have been observed to be inversely proportional to distance from the disposal point.

Disposal of additional Stamford spoil at the N site was also accomplished successfully and a monitoring survey conducted on May 21 indicated the development of a small mound similar to that observed at the S site. At this N site location, 26,000 m^3 of Stamford material were deposited prior to the sand capping.

On June 19, a bathymetric survey was made on the N site to determine any areas that were not covered by sand, and the dredge was directed to dump additional material east of the disposal buoy to insure complete coverage. A final survey was conducted on June 22, after completion of the capping operation of the N site (fig 13).

On June 22, a survey was made of the S site to determine the success of the silt capping operation. The contour chart and the vertical profiles (fig. 14) both indicated a distinct mound had developed at the disposal site with a minimum depth of 16 meters and a thickness of up to 4 meters over the Stamford spoils. Because the silt material from New Haven was cohesive, the resulting spoil mound did not have extensive spreading. Although the vertical profiles indicate all Stamford material was capped, future operations with silt should be designed to spread the capping sediment and reduce the thickness to some extent. The volume of New Haven spoil dumped at the S site was estimated at 76,000 m³ from scow load measurements, of which 72,000 m³ (or 95%) was accounted for by volume-bathymetry calculations.

The associated vertical profiles (figs. 13 and 14) indicate that all Stamford spoils were capped by the sand material. However, since the sand was less cohesive than the silt, it tended to flow during deposition thus creating (as shown by the figure) a broader, flatter mound than that developed by silt at the S site. At the time of the June 22 survey the capping layer had a maximum thickness of 3.5 meters over the Stamford spoil mound. This cap was a smooth blanket of sand that divers were unable to penetrate more than 10–15 centimeters by hand digging. A calculation of the volume of spoil and sand deposited since the May 21 survey indicated an increase of 33,000 m³. This volume compared favorably with dredge volumes specified by the hopper dredge, however, large correction factors based on density and water content of the sand, made comparisons tenuous and calculations of volume and percentage lost to the water column of doubtful validity.

The results of these surveys indicated that the capping procedures employed during the Stamford/New Haven disposal operation were successful. The precision disposal of Stamford spoils resulted in a small compact mound that was readily covered with New Haven material. Apparently, there is little difference in the ability of sand or silt to accomplish the desired capping. In the case of sand, the capping layer is not as thick, but the smooth, dense nature of the deposit acts as a tough, impervious blanket over the capped sediment. Silt deposits on the other hand, derive their capping ability from the cohesive nature of the sediment, developing a thicker deposit with rougher topography.

The ultimate effectiveness of the capping procedures depends on the stability of the resulting cap and its ability to isolate the enriched spoils from the biota and the water column. Consequently, following

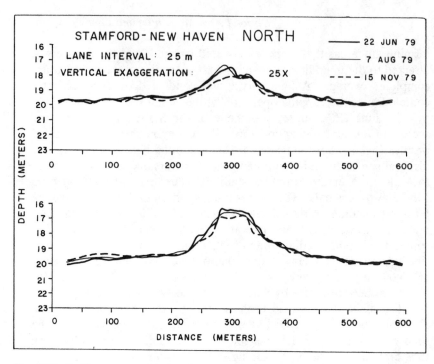

Fig. 9.13 Profile sequence showing changes occurring at North Site from June to November 1979.

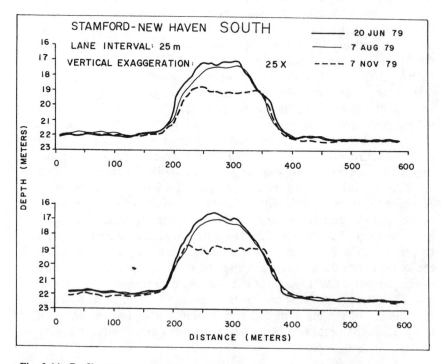

Fig. 9.14 Profile sequence showing changes occurring at South Site from June to November 1979.

deposition of the soils the thrust of the monitoring effort was to evaluate the stability of the resulting mounds with time. Again, this was a multidisciplined effort involving physical, chemical, and biological measurements.

Surveys were made in August, 1979 (figs. 13 and 14) indicating no significant changes in the spoil mounds or the capping material except for a slight settling or consolidation of both spoil mounds. These results were expected since the spoil mound from the 1974 dredging operation has been stable for several years indicating the containment potential of the disposal site.

A second post-disposal survey was conducted on the S site (silt cap) on November 7, 1979. The results showed a surprising change in the topography of the S spoil site equivalent to a loss of approximately 100,000 m³ of spoil from the top of the mound (see figs. 15 and 16). Vertical profiles across the center of the mound (figs. 14 and 16) revealed a flat surface at 19 meters resembling a "guyot" which was also readily apparent on the contour chart. Although this loss of material did not expose any Stamford spoils, further investigations were initiated to determine the causes of spoil movement and to check the other sites.

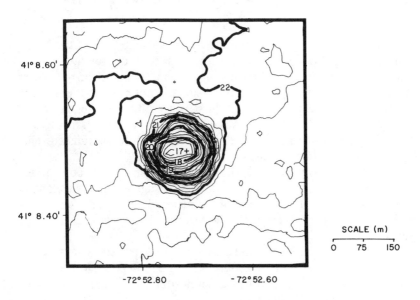

STAMFORD-NEW HAVEN SOUTH, 7 AUG 79

Fig. 9.15 Contour map of the South Site before Hurricane David, 7 Aug. 1979.

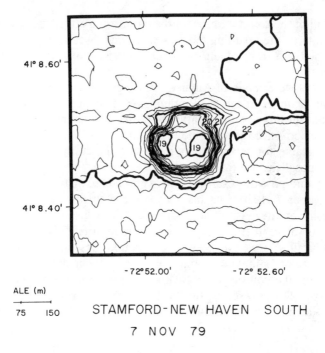

ALE (m)

75   150   STAMFORD-NEW HAVEN   SOUTH

7 NOV 79

Fig. 9.16   Contour map of the South Site after Hurricane David, 7 Nov. 1979.

Additional surveys were conducted on November 15 of the N site (sand cap) (fig. 13) and of the 1974 middle site. Both precision surveys indicated that no significant change in bathymetry had occurred. What dynamical effect was at work at the S site and not present at either the old middle site and in the N site? First, since all three sites lie within 1500 meters of each other and since the bottom is comparatively flat, it is unlikely that one site would experience very different tidal currents however amplified by storms.

What about wave action? The flat topography of the spoil surface which occurs at a depth of about 19.5 meters suggests that wave action was responsible for the removal of the material and a very likely, candidate for supplying high energy waves was Hurricane David which passed through.

We note that the S site (sand cap) occurs in 22 meters depth whereas the N site (silt cap) lies in about 19.5 meters depth. However, the minimum depths for both north and south sites according to the August 7 survey was about 17 meters.

Storm waves present in this area generated by 20 m/s (40 knot) winds might have wavelengths of 150 m and 8 s periods; these would be as "intermediate waves" with flattened elliptical orbital motions at the bottom. Small amplitude theory calculations indicate that such waves at 17 meters (the mean N and S site peak heights) have 20% higher kinetic energy at the 19.5 meter depth (N site) than at the 22.0 meter depth (S site). This is curious since more erosion occurred at the south site. On the other hand, the south site had twice the volume and twice the areal extent for which the waves could interact. Further, the S site also had much steeper slopes which could affect a focussing of energy on the mound. It is possible that these topographical factors could have outweighed the depth effect and brought about a focussing of stress and energy concentration, preferentially upon the S site.

Probably the most important clue is the gross difference in physical and lithological character of the two capping materials. The N Stamford/New Haven and the 1974 New Haven (center) spoil mounds can be distinguished from the S site on the basis of a surface of fine sand material which is probably thicker on the newer spoil mound. This lithology is in sharp contrast to the cohesive silt surface of the southern mound which is characterized by randomly distributed clumps of cohesive clay interspersed within a fine silt matrix.

Calculations were made of theoretical shear stress developed by "hurricane waves" over the rough surface of the south site and compared with stress developed over a smooth surface. These stresses were then compared with estimates of critical shear stress to determine the potential for sediment motion. The calculated shear stress for roughness heights similar to those present on the south site are near or exceed the critical value needed for erosion at wave heights and periods expected from Hurricane David. In contrast, the shear stress developed over the smooth sand surface of the other sites never exceed the critical value. Consequently, it is tempting to conclude that the high roughness factor associated with the clumps of cohesive sediment on the south site create a greater stress and promoted strong sediment motion and a high amount of resuspension under storm wave conditions, while the smoother mounds produced significantly smaller stresses thus insuring the stability of the spoils even at shallower depths.

Preliminary data from subsequent surveys of all sites in April, 1980 show no additional changes to either site indicating that both mounds have since remained stable under nominal tide and wave conditions.

In summary, the effectiveness of capping enriched spoils with cleaner material looks promising, but requires additional monitoring to evaluate long term stability and recolonization effects. The fact that both the sand and silt caps were effective in containing the contaminated spoils during the passage of a "10 year frequency" storm such as Hurricane David is a strong indication of the potential of this procedure.

FUTURE PLANS

Future work is planned to continue concentrated studies of active disposal sites to increase the overall understanding of dredge material dynamics in the marine environment. Additional emphasis is being placed on *in-situ* measurements of turbulence, sediment resuspension, utilization of caged mussels as a field bioassay approach to disposal monitoring and continued visual observations by divers to evaluate long term changes to the dredged material.

It is hoped that the biological studies can be expanded to examine the problem of long term toxicity effects upon the food chains. In short, it is desired to obtain an increasingly more quantitative assessment of the interaction of the dredge spoils upon the environment, a necessity for us to wisely manage safe and effective ocean disposal programs.

References

Bohlen, F., 1980. Suspended sediment transport studies. Proceedings 1979 DAMOS Annual Symposium, Newport, RI. (In press).

Brooks, A., 1980. Benthic ecology, benthic macrofauna. Proceedings 1979 DAMOS Annual Symposium, Newport, RI. (In press).

Environmental Protection Agency, 1977. Ocean dumping act, guidelines for specification of disposal sites for dredged or fill material. Federal Register, September 18, 1979.

Feng, S., 1980. Mussel watch, Second International Ocean Dumping Symposium, Woods Hole, MA. May 1980.

Ianniello, John, 1979. Bolt system progress report 1978–1979. NUSC, New London, CT.

Jones, E., 1980. Sediment chemistry, chemistry of surface sediments in proceedings 1979 DAMOS Annual Symposium, Newport, RI. May 1980.

Massey, A. and Cook, G. S., 1979. Navigation and data acquisition control for the north wall experiment. Science Applications, Inc., Newport, RI.

Morton R., 1980. The management and monitoring of dredge spoil disposal and capping procedures in Central Long Island Sound. Second International Ocean Dumping Symposium, Woods Hole, MA. May 1980.

Pearce, J. B., 1980. The effects of pollution and need for long term monitoring. HELGL Wiss, Meeres. (In press).

Pratt, S., 1980. Fisheries program. Proceedings 1979 DAMOS Annual Symposium, Newport, RI. May 1980.

Shonting, David; Temple, Paul; and Roklan, J., 1979. The Windwave and Turbulence Program, (WAVETOP), Report No. CG-D-68-79. U.S. Coast Guard Office of Research and Development, Washington, D.C.

Shonting, David, and Temple, Paul, 1979. The NUSC Windwave and Turbulence Observation Program (WAVETOP). A Status Report in Marine Forecasting Predictability and Modelling in Ocean Hydrodynamics: 161–182.
Smith, J. D., 1978. Measurement of turbulence in ocean boundary layers. Proceedings of a Working Conference on Current Measurement, Technical Reference DEL-SG-3-78 College of Marine Studies, University of Delaware, Newark, DE.
Stewart, L., 1980. Diver observations. Second International Ocean Dumping Symposium, Woods Hole, MA. May 1980.
U.S. Army Corps of Engineers, 1977–1978. Dredged Material Research Program, Series of Reports, Waterways Experiment Station, Vicksburg, MS.
Walker H.; Saila, S.; and Anderson, E., 1979. Exploring data structure of New York Bight, Benthic Data using post-collection stratifications of samples and Linear Descrimination Analysis for Species Composition, Estuarine, and Coastal Sciences, No. 8.

Appendix: DAMOS Bibliography

Damos Contribution Reports

Stamford/New Haven disposal operation monitoring survey report-baseline surveys. NED, Army Corps of Engineers. SAI/DAMOS Contribution #1. March 30, 1979.
Stamford/New Haven disposal operation monitoring survey report–20,000 yd³ Increment. SAI/DAMOS Contribution #2. April 13, 1979.
Stamford/New Haven disposal operation monitoring survey report–50,000 yd³ southern site, 10,000 yd³ northern site. SAI/DAMOS Contribution #3. April 30, 1979.
Completion of Stamford disposal (Stamford/New Haven). SAI/DAMOS Contribution #4. April 1979.
Post disposal surveys (Stamford/New Haven). SAI/DAMOS Contribution #5. June 3, 1979.
Post disposal monitoring (Stamford/New Haven). SAI/DAMOS Contribution #6. August 1979.
Stamford/New Haven disposal operation monitoring survey report. SAI/DAMOS Contribution #7. January 30, 1980, 40 pp.
Management and monitoring of dredge spoil disposal and capping procedures in Central Long Island Sound. SAI/DAMOS Contribution #8. April 16, 1980.
Chronological records of in-situ physical and biological conditions obtained by diver survey at New London and New Haven Connecticut dredge disposal sites. SAI/DAMOS Contribution #9. Second International Disposal Conference (In press).
Changes in the levels of PCBs in *Mytilus Edulis* associated with dredge spoil material. SAI/DAMOS Contribution #10. Proceedings Second Annual

Ocean Dumping Symposium April 1980, Woods Hole, MA. Contributions 8, 9, 10.

"Capping" procedures as an alternative technique to isolate contaminated dredge material in the marine environment. SAI/DAMOS Contribution #11. Testimony before U.S. House of Representatives, Committee on Merchant Marine and Fisheries. May 21, 1980.

Precision disposal operations using a computerized Loran-C System. SAI/DAMOS Contribution #12. May 1980.

Disposal area monitoring system, progress report. SAI/DAMOS Contribution #13. March 15, 1980–May 1980.

Disposal area monitoring system, progress report. SAI/DAMOS Contribution #14. May 15, 1980–July 30, 1980.

Miscellaneous DAMOS Reports

Disposal area monitoring system–annual data report, 1978. Technical Publication, New England Division Corps of Engineers, Waltham, MA. May 1979.

Disposal area monitoring system–annual data report, 1980. Technical Publication, New England Division Corps of Engineers, Waltham, MA. (In press).

Cook, G. S., R. W. Morton and A. T. Massey, A report on environmental studies of dredge spoil disposal sites: Part I–An investigation of a dredge spoil disposal site; Part II–development and use of a bottom boundary layer probe, in: *Bottom Turbulence.* J. C. J. Nihoul, Ed., Elsevier Oceanographic Series #19, 1977.

Baseline environmental measurements at the proposed portland disposal site, NUSC T.R. (in press) – Manuscript at NE Division, Corps of Engineers.

Disposal area monitoring system FY-77 progress report, NUSC T.R. (in press) Manuscript at NE Division Corps of Engineers.

Preliminary site survey, MK 48 FOT&E program shallow water test, Gulf of Maine, NUSC TM# 311-4352-76(c). Naval Underwater Systems Center, Newport, RI.

Bottom current measurements from the Long Sand Shoal dumping ground NUSC TM# TA132-4002-75. Navel Underwater Systems Center, Newport, RI.

A summary of environmental data obtained at the New London dump site and the East Hole, Block Island Sound. Final Supplement to EIS Thames River Dredging Project. Northern Division, Naval Facilities Engineering Command. 1979.

A preliminary of current and bathymetry measurements of alternate disposal sites for the Thames River dredging project NUSC TM# TA132-4578-74. Naval Underwater Systems Center, Newport, RI.

Summary of current data from the acid barge site south of Block Island NUSC TM# TA132-4182-74. Naval Underwater Systems Center, Newport, RI.

CHAPTER 10

Chronological Records Obtained by Diver Survey at the New London and New Haven Dredge Disposal Sites

LANCE L. STEWART
University of Connecticut
United States

INTRODUCTION

One of the original motivating reasons for our involvement with dredge disposal monitoring was to try to understand how such disposal affects the biota, and whether the catastrophic impacts predicted when these operations first began have materialized.

Monitoring efforts began in the mid 1970s, when environmental concern was at its peak and we knew very little about what the effects in the ocean were going to be.

It is necessary to consider positive aspects or advantages of intentional dredge disposal pollution. First, from a managerial sense, the process can be *timed*. Timing is very important in a biological world because of seasonal cycles of abundance and reproductive activity. It also is very important due to potential commercial fishing impact. For example, the New Haven project in Long Island Sound is in the vicinity of large scale oyster operations, and the dredge sites are critical spawning and spat fall grounds. If dredging is permitted during a larval period or a spat fall period, then inhibition by siltation could be extreme and destroy an entire year class of oysters.

Timing in this particular case was accomplished by the DAMOS Program and several different agencies and groups (N.O.A.A., Fish and Wildlife Service, the Corps, and fishing industry representatives) through compromise discussions to determine critical times and areas to avoid.

A second rationale in intentional pollution of the ocean, would be advance determination of the *pollutant characteristics* and the degree of pollution of the material that you intend to dump. Connecticut has determined three criteria levels for pollutant load in sediments, and prescribed methods of disposal for each of the three categories.

A third important logistic to consider would be *site selection*. Selection of a containment site by pre-survey, to retain materials on the bottom in a limited location, would avoid destruction of productive habitats through a more refined site selection process.

As a fourth point, a logical and systematic *long-range monitoring* program should be designed to detect dredge disposal impacts. In a large-scale experiment of this nature, direct field observation and interpretation are essential to test the merit of certain site-specific disposal techniques. Alternatives in disposal containment also may be investigated. For example, Dave Shonting, in Chapter 9, cited capping as an apparently effective method of disposing of contaminated material. There are other methods such as burrow pits, which offer even a greater measure of protection by placing material below the horizon of transport.

The DAMOS Program represents a multi-disciplined assessment of environmental impact caused by dredged sediment and chemical constituents placed in various ocean environments. Specific areas of investigation include: precision bathymetry, sediment chemistry, hydrographic water parameters, sediment transport mechanisms, benthic infauna analysis, heavy metal and PCB bioaccumulation, fishery survey, in-situ inspection and photographic assessment.

Figure 2 in Chapter 9 indicates New England area disposal sites under survey. In the confined region of Long Island Sound, state jurisdiction extends to the mid-Sound, Connecticut-New York border. The active sites in southern New England, two miles south of New London and seven miles south of New Haven, have drawn controversial environmental concern, a high public interest level and intense supervision by E.P.A. and other regulatory agencies.

With those considerations of intentional pollution in mind, we will proceed with a short discussion of our *in-situ* approaches to the monitoring program.

Chronological records of biological and physical benthic conditions have been obtained for the New London, Connecticut ocean disposal site since June 1974, and for the New Haven site since March 1979. Extensive dive investigations have been conducted during preactive, and post disposal phases. Certain methods and procedures have been effective in understanding spoil sediment dynamics and assessing impact and faunal behavior at dredge disposal sites. Underwater photodocumentation, discrete diver sampling, and periodic transect observation are valuable techniques incorporated in disposal site evaluation. The following is an outline of basic *in-situ* survey objectives for ocean disposal site evaluation.

1. *Preliminary underwater survey of pre-disposal sites to determine:*

   Occurrence and location of habitats supporting populations of important economic species (mollusks, crustaceans, and benthic finfish);
   Baseline characteristics of fauna and habitat associations;
   Identification and enumeration of major macrofauna by transect or video tape survey:
   Species composition of macro-benthos with particular attention to degree of biological community diversity.

   Bottom sediment type, surface features, and topography of marine terrain (i.e., ripple marks, shell layers, burrow excavations, sediment compaction and homogeneity).

2. *Active dredge disposal observation to detect:*

   Characteristics of dredged material on impact and placement at disposal site, degree of visual turbidity, unique sediment transport processes, uniformity of dredge material type and limits of dispersion.

   Biological recolonization, identification of opportunistic species, community structure developing over time, invasion trends, areas of concentration;
   Discrete sampling by diver operated epibenthic net, benthic plot, core sampler, and penetrometer

   Faunal behavior in spoil vicinity — reaction of major commercial species to spoil material (e.g., attraction, avoidance, smothering, transplantation, apparent toxicity, bioturbation or stabilization).

   Operations assessment: capping procedure, degree of containment, dumping accuracy (implications of misplaced load and effect on bio-accumulation experiments), proximity of spoil to establishing sampling stations.

3. *Post disposal site monitoring to obtain:*

   Complete record of recolonization, succession and bottom feature changes over the long term;
   Permanent sampling and photographic stations at spoil border via sonic beacon marker, transect orientation line, calibrated elevation stakes, and penetrometer test sites.

   Circumnavigation of spoil periphery coordinated with surface tracking to delineate spoil border and area of coverage;

Stability of spoil frontal boundaries by permanent sonic markers and correlation of this method of charting spoil circumference with surface bathymetry.

IN-SITU INVESTIGATIONS AT THE NEW LONDON OCEAN DISPOSAL SITE

Underwater observation and photography was conducted in the vicinity of the New London, Connecticut disposal site throughout all seasons of the year. Dives were made at selected stations on the disposal site to determine typical macrobenthos and sediment characteristics within the area. Description of biological activity focused on commercially important species and recolonizing species inhabiting spoil material. Station transects were covered at approximate one to three month intervals with all notable observations recorded by 35-mm still photography. Average bottom time for transects was twenty-five minutes, compass courses followed were perpendicular or oblique to bottom current and distance ranged from 50–300 meters. In addition to the periodic visual documentation, data collected on each dive included: depth, tidal stage, bottom visibility, estimate of current velocity and direction, bottom type, and predominant benthic species.

Predisposal surveys for the New London region revealed disposal site benthic conditions consisting of featureless, flat sediment heavily populated with Ampeliscid (amphipod) communities. After initial disposal, scour and slight surface erosion was evident on phase I spoil at the New London buoy. The large cohesive clay masses, previously protruding one to two meters from surrounding sediment had flattened considerably after a three to four month period.

Ten to twenty meter diameter patch areas of extremely different sediment grain size, varying from depressions containing soft silt to the rarer (5–10%) high spots of compact gravel, were observed in an easterly transect from the New London buoy location. Extensive shell surface layers occurred along the transect, and the capping material therefore provided an attachment substrate for fouling organisms.

The characteristic cohesive clay material was often multi-faceted and fissured on impact with the bottom forming invasion points for epibenthic species. Immediately following first disposal from phase II, the existence of small (0.5–1.0 centimeter) clay balls were noted on the spoil surface. These spherical fragments appeared to vary in density and collect to the lee of large clumps. The "clay balls", moved by relatively low current velocity, were observed to roll over spoil surface and cascade down slight slopes. The implication of a new sediment transport mechanism, other than by resuspension velocity/grain size theory, is noteworthy here. The degree to which this process occurs and the causes of formation, either biotic or physical, are under investigation.

Unexpected underwater location of spoil material was charted in certain cases and found to be outside the normal dumping pattern. These isolated spoil piles were probably the result of errant barge release. All were inside the designated disposal site, but in relation to experimental design of the sampling process, could produce complications in interpreting both bottom grab (species diversity) and shellfish (heavy metal) data.

A winnowing of the fine surface material has occurred and ripple-marks (two centimeters amplitude and 10–15 centimeters frequency) were observed at points in the direct vicinity of the New London buoy in August 1977. A herringbone pattern twenty to fifty meter diameter areas resulted from collection of detrital material in the troughs. Noticeably more organic material and peat masses comprised dredged material deposited from phase II operation.

Diver visibility has seldom been limited by spoil resuspension throughout the course of underwater surveys. The only incidents of spoil turbidity occurred when investigating an area within one day of disposal or when immediately downstream from spoil disposal and the spoil plume passed overhead in the water column. No visual observations of sediment-water interface turbidity produced by normal tidal currents up to one knot have been made. Storm induced turbidities account for greatly reduced visibility. The beneficial factors of increased oxygenation of anoxic sediments compared to detrimental factors of greater water contact/resuspension and faunal contact with contaminated material must be considered. No areas of sediment discoloration due to generation of hydrogen sulfide or sulfide bacterial films have been observed on the New London disposal site. An overview of the various visible organisms responsible for bioturbation on the New London disposal site was obtained by direct observation on over fifty successive dives to date. A few qualitative descriptions of the species and behavioral traits, accounting for disturbance of recently deposited spoil, are listed in table 1.

The high degree of biological activity at the sediment-water interface of the New London site results in the new spoil material being continually excavated, manipulated, and fragmented resulting in changing spoil profiles. Principal locations of activity are at the spoil pile perimeter or in and around the base of clay mounds. This activity is not restricted to the top few centimeters of sediment but may extend, in case of decapod crustacea, to depths up to one meter.

TABLE 10.1

SPECIES AND BEHAVIORAL ACTIVITY CONTRIBUTING TO MAJOR BIOTURBATION OF SPOIL SEDIMENTS ON THE NEW LONDON SITE[*]

SPECIES	SEASON PRESENT	STEREOTYPED BEHAVIOR	RELATIVE ABUNDANCE
Busycon canaliculatum (whelk)	Spring–summer	Burrow beneath substrate, predator seeking small bivalve mollusks	2–3 per dive, often undetectable buried in sediment.
Polinices heros (moon snail)	Spring	Substrate burrower	Rare
Asterias forbesi (starfish)	Year round	Dislodge mollusk infauna	Common summer, 6+ per dive
Cancer borealis (Jonah crab)	Spring–fall	Excavation of shallow depression and pits, sediment fragmentation of clay banks	Common 6+ per dive, associated with cohesive clay masses
Cancer irroratus (rock crab)	Spring–fall	Settle into soft substrate to carapace level — resting posture, protection	Rare, few at periphery of spoil
Libinia emarginata (spider crab)	Fall	Burial of entire carapace and legs.	Occasional, in aggregations
Homarus americanus (lobster)	Year round	Extensive excavation to 1 m deep, expansion of *C. borealis* holes to large dimensions	Common 2–6 per dive
Pagurus bernhardus (hermit crab)	Spring	Feeding behavior of thrusting chelipeds into soft sediment to seize infauna	Occasional
Pagurus longicarpus	Spring	Manipulation of surface shell fragments for food detection	Occasional
Prionotus carolinus (sea robin)	Summer	Use of modified pectoral rays to seek food by manipulating surface layer of spoil	Common summer, 2–3 per dive
Urophysus chuss (hake)	Spring	Burial of caudal portion of body in soft substrate	Occasional
Pseudopleuronectes americanus (flounder)	Spring–summer	Fanning behavior to expose infauna in selective feeding; self burial utilizing dorsal and anal fins	Extremely abundant, 6–20 per dive
Paralichthys dentatus (fluke)	Summer	Self burial as predator ambush posture	Rare summer
Myoxocephalus ortodecemspinosus (sculpin)	Winter–spring	Settle partially into soft substrate during resting state.	Common winter

[*]Noted in repeated survey dives in the area

The relief of the spoil piles added to a surrounding flat bottom has been noted to attract a wide range of benthic species, including several benthic finfish (i.e., flounder, scup, hake, sculpin, and sea robins) (figure 4). The eddy conditions to the lee of mounds concentrate detritus and establish areas for intensive feeding. In several cases mollusks and associated infauna damaged by successive disposal were observed to create a "chumming effect." The common bivalve predators, *Asterias* (starfish) and *Busycon* (whelk), are often present on new spoil.

Bioturbation, the physical and chemical effect of the biota on their environment, is of great importance to dredge spoil surface dynamics. Mobility of the major epibenthic species does not appear to be inhibited by sediment conditions in and around recent spoil. A consequence of the soft sediment facilitates the grazing and substrate feeding behavior.

There appears to be a strong recolonizing force at the New London disposal site, but the rapid repopulation rate has been inhibited by sporadic dumping schedules. A suggestion is made that large scale dredge/disposal operations be continuous, because cessation periods of one to two months allow a high degree of repopulation of organisms vulnerable to smothering once disposal recommences. Resources of particular economic importance affected in this manner include the lobster, cancer crabs, and mollusks.

Throughout the dredge disposal operation there has been mass transport of the bivalves, *Mercenaria mercenaria* and *Pitar morrhuana* (as well as other estuarine infauna) from the Thames River to the New London disposal site. Evidence of individuals deposited on or near the surface indicates survival. The fate of more deeply buried individuals is unknown. However, dive observations reveal there is a good possibility that substantial numbers of commercially important species (the quahog, *Mercenaria mercenaria*) may have been introduced to the previously unproductive New London site. Quahogs undergoing this transplant and remaining on site may be able to be depurate toxic compounds and pathogens and be fit for harvest and consumption. The long-term survival and eventual fate of this population should be followed.

The occurrence of several unusual and opportunistic species have been recorded in the New London vicinity. Seasonality appears to play a key role in influencing the abundance of species on new spoil material. At times, numerous "pollution intolerant" species were found indicating no apparent repellent effects or high sensitivity to the spoil. Dives in the spring of 1977, revealed dense concentrations of the solitary stalked hydroid, *Corymorpha pendula*. The apparent bloom of this species produced densities up to 15 m² during May 1977, with gradual disappearance from the area in June to August. Other typical offshore or "pristine water" species included the northern shrimp, *Dichelopandalus* and scal-

Fig. 10.1 North-northwest of New London: The solitary hydroid, *Corymorpha pendula.* Occurrence noted on virgin phase II spoil in dense concentrations up to 10–15 m², spring of both 1977 and 1978.

lop, *Pecten magellanicus.* Shrimp were abundant in winter dives and were found associated with new mussel beds in the south and southwest disposal sectors. Only occasional sea scallops have been sighted.

One notable observation was the lack of obvious amphipod recolonization up to May 1979. The dense tube mats of ampeliscid communities previously characteristic to areas in the northeast and southeast sectors had not been encountered on or directly adjacent to the spoil pile. In Rhode Island Sound disposal investigations, it has been suggested that colonization of a spoil mound by such tube building animals could be expected to slow the rate of erosion. This did not occur at the

New London site during the first three years of monitoring, but rather other species account for the stablizing effects noted (table 2).

The reappearance of tubicuous amphipod communities (*Ampelisca* sp.) was first noted on phase III peripheral regions in June 1979, and later in the fall near the phase IV active disposal area. Photographic estimates of visible parchment tubes on spoil sediment surfaces approach densities of 9000/m^2 (Fig. 10.2). The recurrences of these populations are now comparable to amphipod community densities observed in pre-disposal reconnaissance of the NE sector in June 1974.

Associated diver observations in June and July 1979 showed evidence of extensive burrow formation by *Homarus, Cancer* and *Urophysis* at NW, SW and SE peripheral stations. Spawning of large gastropod snails (*Polinices, Lunatia, Busycon*) occurred at several disposal pile locations as evidenced by diver count of abundant circular egg collars (1–3/m^2) in June 1979. Again, the obvious attraction to clay mound topography and 20°–30° slope contours on the disposal pile resulted in high diver species count for: winter flounder, *Pseudopleuronectes americanus* and Squirrel hake, *Urophysis chuss,* the predominant benthic fish.

Sequential Bioturbation

The succession rate (resilience of communities) has been shown to be slower in the deeper, less frequently disturbed environment than in the inshore shallow water environment prone to storm disruptions. However, differences between the types of organisms responsible for sediment reworking (bioturbation) in deeper water disposal sites must be recognized. The smaller infaunal community assemblage requires considerable time for repopulation and sediment bioturbation attributed to this group is a slow process. The sediment turnover depth has been postulated to gradually increase to a ten centimeter maximum due to this type of recolonization. In contrast, initial bioturbation of newly deposited dredged spoil observed for mobil macrobenthic species is more rapid and is postulated to be a major mechanism for "conditioning" surface sediment layers for later patterns of infaunal succession. The macrobenthos, (including benthic finfish, large crustacea, predaceous mollusks, and motile echinoderms) may be characterized by: immediate invasion tendency, significant biomass, transient nature, active sediment manipulation for feeding or shelter, and turnover effects to depths exceeding one meter. Immigration of predominant species from surrounding communities is facilitated by this process. Numerous excavation areas, attributed primarily to transient macrobenthos, are es-

TABLE 10.2

SPECIES AND MODE OF ACTION CONTRIBUTING TO STABILIZATION
OF NEWLY DEPOSITED SPOIL OBSERVED AT THE NEW LONDON SITE

SPECIES	RECOLONIZING PATTERN	RELATIVE ABUNDANCE
Sponge		
Halichondria sp.	Encrusting shell fragments and attaching to calcareous substrate	Rare
Cliona sp.		Small colonies
Coelenterates		
Tubularia sp.	Individual tufts on exposed rock or shell	On older spoil, hard surfaces
Corymorpha pendula	Opportunistic, stalk with substrate holdfast	Bloom in spring 15m^2 density
Cerianthus americanus	Large subsurface tube	Rare
Mollusks		
Mytilus edulis	Massive beds formed in interconnected net pattern, byssus thread attachment to coarse surface material	Extensive coverage in areas of set
Polychaets		
Diopatra cuprea	Large agglutenated tube, solitary	Occasional
Capitella sp.	Patchy concentrations of several individuals	Tube densities up to 30 m^2
Nephtys sp.	Clusters of tubes in clay mounds	recorded

Fig. 10.2 Southeast transect from the Northwest Border Station: Adjacent burrows (*Homarus* and *Cancer*) excavated in clay banks on the spoil surface demonstrate substrate cohesiveness. Extensive amphipod communities (tubes protruding from sediments) have repopulated the SE and NW spoil sectors.

pecially concentrated in a twenty meter peripheral band around the entire spoil pile.

Sediment Features

Extreme surface grain size heterogeneity has been reported for the New London disposal site, with alternating ten to thirty meter patches of soft-clay mud to coarse gravel sediment. The infaunal associations that can be expected to settle, according to classical organism/sediment grain size relationship, would be extremely variable over the mosaic surface.

This fact is cited as probable cause of complications in species diversity index studies and predictable succession series with time for any disposal site.

Fishing Activity and Gear Effect

In addition to the magnitude of sediment surface "conditioning" by mobil macrobenthos, the artificial effect of trawl gear on spoil pile disturbance can be great. The sediment disruptions occurring during passage of otter doors and a weighted sweepline can, at least, be termed severe in relation to small infaunal effects. The use of commercial trawl gear has been noted in the New London disposal vicinity during 1978. Furrows in the spoil surface one meter deep have been observed on other sites, and the consequences of trawl operations, re-exposing contaminated buried fields, should be considered in future management of disposal sites.

Natural Mussel Population

First evidence of major epifaunal spoil coverage was found at a point .5 N miles south of the New London buoy in September 1977. Periodic diver collection of *Mytilus edulis* in the southwest sector location, has indicated progressive growth of the population throughout the sample period, September 1977–November 1978. First observation, September 22, 1977, revealed .5–1 centimeter shell length individuals distributed over approximately 25% of phase I surface in a net growth pattern. Setting presumably occurred in July 1977. Population samples on December 16, 1977 demonstrated mean shell length increase to 2 centimeters and an increase in percentage surface coverage to 75%. Subsamples of the mussel population March 29, 1978, with one hundred *Mytilus* shell length measures gave a 3.3 centimeter mean. The final sample and measurement of one hundred live *Mytilus* shells, November 15, 1978, gave a 4.1 centimeter mean length. Health of the bed appeared good throughout the 1978 sample period and intensive predation was noted (Fig. 10.3). Percentage of phase I spoil coverage remains fairly consistent at 50%, and banks of mature mussels are forming over an extensive area in the southwest sector.

New phase IV spoil deposited in the southwest sector region September 1979 prevented continual surveillance of the population longevity. However, February 1980 sample data reflects an occurrence of smaller shell length indicating additional set and growth had occurred in the vicinity. Diver survey of the southwest region, after new spoil obliterated the southwest benthic border station, revealed a dense, ju-

Fig. 10.3 Southwest border station: *Mytilus bysuss* thread attachment to shell fragments solidifies spoil surface material, and interspace between mussels acts as sediment trap. Intensive predation by *Asterias* was noted on phase I mussel bed.

venile *Mytilus* set atop recent phase IV spoil in the region. This observation of "O" year class (.5–1 centimeter) shell length mussels suggest
the massive repopulation by *Mytilus* in this region is a recurrent process
despite disposal irregularity. Subsamples of diver collected mussels have
been fixed and retained for heavy metal analysis since these *Mytilus*
samples have grown in close proximity to the substrate interface.

Spoil Boundary Areas

On adjacent natural bottom, within ten meters of the phase II spoil
extremity, benthic faunal composition appeared normal. No evidence
of epibenthic smothering was noted to occur far beyond the border
stations as determined by visual inspection. The natural sediment veneer, in flux with each tidal cycle, did not show a wide gradient band
(overburden with spoil material) at border locations in most sectors.
The transition zone from spoil coverage to natural bottom averaged
three to five meters around south, west, and north sectors. Eastern and
southeastern sectors revealed a twenty to thirty meter indistinct transition zone with patch areas of natural bottom enclosed by this spoil
overlay. Adjacent undisturbed bottom communities demonstrated good
diversity with a full range of fouling organisms present.

Immediately following termination of phase II disposal, intensive
dive surveys were conducted at several points around the circumference
of the spoil pile to determine the final boundary location. Divers followed transects perpendicular to the spoil periphery to locate the limit
of spoil dispersion that could be visually detected on the adjacent ocean
bottom. Surface buoys were towed on short tether lines and a pipe
anchor was placed at the point where natural bottom first revealed
evidence of spoil apron coverage. Pipe anchor surface buoys were utilized to assure that if surface buoys were molested or accidently fouled,
the marker would be carried away by tidal action and not replaced in
an erroneous border position. In conjunction with bathymetric survey,
Loran C coordinates for marker buoys were interfaced with the Decca
navigation system and a plot, in relation to the one mile square New
London disposal area was obtained.

Permanent border stations were selected and marked with sonic
beacons in the northwest, southwest and southeast sectors. Underwater
sonic receivers permitted the diver to return to border stations for repeated observation of biological recolonization and spoil border stability. At each station a fifty meter polypropylene line was staked to the
bottom perpendicular to the spoil periphery. Along the line three calibrated (centimeter) .5 meter stakes were driven into the spoil material

at three meter intervals. Repeated (Fig. 4) readings of these stakes were taken to provide a gauge of spoil compaction or erosion.

Photography of vertical sections of sediment has been conducted at transect stations to detect layer/strata conditions occurring in the bioturbation-stabilization process.

The slope and features of the apron at different positions around the spoil pile were noted in the process of marker placement. Along the west, northwest and northeast border a sharp boundary of spoil/natural bottom was evident. The slope of the spoil material was steep (about 20° – 30°) at many points around the west to northwest periphery indicative of individual barge loads with one to two meter banks of spoil above the natural bottom.

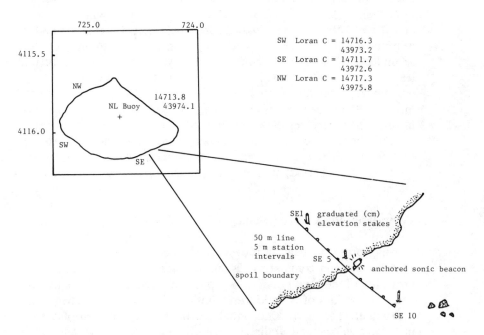

Fig. 10.4 Benthic perimeter station, showing line, staked to bottom, along which stakes are driven into spoil material.

Buoy placement in the eastern to southern sectors revealed an extremely flat spoil apron with an irregular scalloped border. In the border vicinity spoil overlay was thin and graded uniformly from one meter to the one centimeter detectable border over a 100 meter distance. Determination of a distinct spoil border from bathymetric data would be difficult in this sector.

Effects of Storm Events

Inspection at the disposal buoy (March 25, 80) produced unique observations on recent spoil dynamics under post-storm conditions (1–2 meter swell–50 meter frequency). In three separate areas, "fissure lines" (4–5 meter) were observed on the spoil surface. The effect of overhead swell at the fifty foot depth was evident by a .5–1 centimeter movement of opposite sides of the fissure line. Masses of spoil (estimated 50 meter areas) were observed to separate, creating a 1 centimeter wide 30–50 centimeter deep fissure line, and merge together repeatedly on a 7–10 second cycle. No extreme surface turbidity was noted due to current transport however, a 10–20 centimeter vertical extrusion of turbid silt was emitted from the fissure line on each closure. Another type of spoil mass oscillation can be described as a "slip line" with one density spoil material rising and residing on a 30°–45° plane over adjacent material. No turbidity resulted from this sediment action. The energy exerted on the spoil volume produced no translatory effect, during the fifteen minute observation period, but appeared to function as a compaction mechanism.

Commercial-Recreational Fishing

During border delineation procedures, June–September 1978, considerable sport and commercial fishing activity was noted in immediate proximity of the New London spoil pile. The head boats, "Blackhawk" and "Mijoy" out of Niantic were consistently seen with parties of approximately twenty-five individuals on board. The pattern of activity involved a 500 meter drift directly in line along the southern and western disposal site phase II border. These commercial vessels were often accompanied by five to ten small private sportfishing boats following similar fishing procedures.

The attraction of fishing efforts to this area substantiated in-situ observations of abundant winter flounder, *Pseudopleuronectes americanus,* schools concentrated along the entire spoil/natural bottom periphery. In last summer, August–September, the target species shifted to scup, *Stenotomus chrysops.* Catches also included other benthic species: fluke, *Paralichthes dentatus;* tautog, *Tautoga onitis;* and sea robin, *Prionotus carolinus.*

STAMFORD/NEW HAVEN OCEAN DISPOSAL SITE
MONITORING – CENTRAL LONG ISLAND SOUND

In-situ investigations at the New Haven site (both primary south and secondary north target areas) commenced in March 1979. Initial monitoring efforts involved baseline survey and installation of diver transect cables with station markers. Dive investigation objectives were to: determine characteristics of spoil disposal by visual observation; document conditions by underwater photography; conduct systematic sampling; install diver orientation system and station locations (at several stages); delineate visible spoil boundaries; evaluate capping procedure; investigate biological sediment reworking, faunal behavior, and recolonization.

Benthic Transects and Stations

Baseline orientation cables (200 meter and 400 meter lengths) were laid east-west at the respective north and south sites March 22–23, 1979 (fig. 5). Station markers consisted of coded polypropylene line fastened to the cable in logarithmic intervals from the center. Underwater photographs were taken on both sites along the transect cables to document pre-disposal sediment features, benthic organisms, core sample procedures, and calibrated stake placement.

Core Samples

Baseline core samples were obtained from eight selected stations along S and N transect cables. The core sample device consisted of three 7 cm diameter x 20 cm plastic cylinders secured in line approximately 2 cm apart. Six plastic caps were cemented to short lines for closure of each core bottom immediately after sample. Diver collected core samples allowed discrete sampling of the soft sediment with exact reference to station location and benthic topography. The diver samples were to augment surface sediment samples taken by shipboard Smith-McIntyre grab at multiple stations radiating from the central spoil target site.

One meter calibrated elevation stakes were placed along the transect cable at seven of the stations, leaving forty centimeters of calibrated stake exposed above the baseline sediment surface to detect depth of spoil coverage after initial disposal.

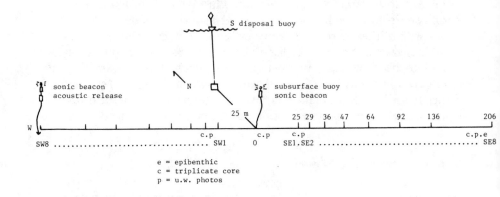

e = epibenthic
c = triplicate core
p = u.w. photos

Secondary (North) Site - baseline transect cable and stations

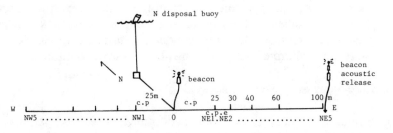

Fig. 10.5 New Haven/Stamford ocean disposal sites, north and south, showing monitoring equipment.

An epibenthic net sample was obtained at selected stations. The .5 m x 20 cm (1 mm mesh) net was diver operated over a thirty second transect of the bottom. Collection by this method allows a standard timed course or known distance to be sampled. The net attitude was adjusted to fish properly over extremely soft sediment and over variable contours. A base bar depth of approximately 1–2 cm beneath the sediment surface was maintained throughout transect sampling.

The basic objective in diver operated epibenthic net collection was to discern differences in occurrence of epibenthic fauna that may not be detectable by Smith-McIntyre infauna sampling. The mobil and transient species are more vulnerable to capture utilizing this method of standard transect length and controlled fishing attitude. A more accurate representation of species occurrence on dredge spoil material is obtained by combining benthic infauna, diver epibenthic collection and visual observations.

On April 10, 1979, a dive inspection of the interim disposal condition on the south site was conducted. Underwater photography and observations on the condition of the spoil surface were designed to: survey the extent of dredged material dispersion north to south; inspect spoil surface characteristics, homogeneity, compaction, topography; and note biological activity on newly deposited spoil.

Detection of spoil limits was possible by observation of cohesive clay mounds, slight textural and color difference, and principally by burial of ubiquitous solitary hydroid, *Corymorpha pendula*. Colonies covering adjacent natural bottom approached average densities of approximately 30/m².

Distribution of spoil material in the target area appeared to be the result of sequential dumping procedures and not the result of dispersion due to current transport. The spoil material observed on the bottom indicated cohesive clay masses with loosely consolidated interspace areas. This first phase material inspected in transect dives appeared stable and not prone to migration. A distinct north and south perimeter could be detected. In general, a majority of the spoil material appeared more compact and of higher density than the surrounding natural bottom sediment.

The normal flux of bottom surface silt veneer overlays approximately three meters of spoil in the apron regions. Tracks from mobil invertebrates were evident on spoil surface and the fragmentation and excavation of cohesive clay clumps had occurred. A high degree of thigomatactic response was exhibited by organisms immediately inhabiting spoil habitat. In general, a greater representation and assemblage of megabenthic organisms was noted on and within spoil material than had been observed in the baseline survey of this area on March 22, 1979.

In April 1979, in-situ observations were made for post-Stamford phase at the south site and interim-Stamford phase at the north site prior to capping. At the south site buoy a post-Stamford disposal pile elevation line was staked extending 30 meters to the northern border extremity and 170 meters towards (within 10 meters) the southern border. The line provides a reference for New Haven capping effectiveness.

A dive survey of northern and southern sites in June investigated the final capping procedure for effective coverage, characteristics of sand vs. silt-mud overlay, apparent fluidity or mobility of final dredge deposits, and differentiation of spoil boundaries at both sites. Considerable debris was observed to be incorporated in spoil material (i.e., steel bulkheads, pipe, rope, sheet metal, plastics, bottles, cans, etc.).

Divers performed a cap coverage survey on the north site at locations where bathymetry indicated thin overlay. This condition was confirmed and the Army Corps of Engineers vessel *Essayons* added New Haven sand to achieve greater cap depth. Observations on the sand cap characteristic indicated a dense "resilient" sand layer at the spoil pole flanks. On excavation the coarse sand would flow to fill in surface furrows.

South site observations indicated a difficulty in distinguishing silt-mud cap overlay sediment type from similar Stamford material. The characteristics of numerous clay mounds with angular facets and extremely irregular topography was used as indication of cap overlay. The north-south ground line, placed over the final Stamford central mound disposal (April 1979), was not found on a diver search and further confirmed cap coverage. At the spoil periphery evidence of new faunal activity (burrows, cones, mollusk trails, clay base excavations) demonstrated spoil material along the entire diver transect was of recent New Haven origin. At the final cap stage (June 79) for both north and south sites, a diver orientation cable was laid east–west, identical in length and station interval to the original baseline orientation cables. All monitoring reference to cable stations from June '79 on, describes permanent locations over the final New Haven/Stamford capped surface.

The stalked hydroid, *Corymorpha pendula,* was no longer present and available as an indicator species during the summer season at north and south locations. A burrowing anemone, *Cerianthus americanus* was present on natural bottom in sufficient densities to distinguish spoil boundaries at most sectors. Numerous vertical burrows were sighted along north and south transect cables with densities up to 6/m². The majority of these burrows were assumed to be constructed by *Axius serratus* (two live specimens collected on June 19) but burrows formed by *Squilla empusa* can not be discounted.

Benthic feeding finfish (*Prionotus carolinus, Raja* sp., *Scophthalmus aquosus, Pseudopleuronectes americanus, Urophysis* sp.) were most commonly of juvenile size range (five to twenty centimeters). These species were sighted consistently, but their occurrence did not suggest dense concentrations or distinct areas of congregation.

The post capping survey (July of 1979) included standard faunal and sediment feature observations with super 8-mm cinema film sequences along both north and south orientation cables. Emphasis was directed to placement of calibrated elevation stakes at station locations and preliminary penetrometer tests. Readings from previously placed elevation stakes that showed no detectable erosion at station locations. Epibenthic net samples and penetrometer measurements were obtained along north and south pile transects and sediment profile photography was conducted at selected sites on and off disposal piles.

The south site had received new spoil material (November 1979) which overlay the diver orientation cable and prevented continuation of permanent station, elevation stake post-disposal monitoring. Inspection of new spoil topographic relief and limits of coverage was accomplished in free transect courses radiating from the south site buoy. A rapid smoothing of the spoil surface occurred at the south site and recent observations revealed fractured clay mounds, extensive benthic burrowing and dense concentrations of polychaete worms and mysid populations over the spoil surface.

The north site transect cable has remained in place and has allowed repetitive inspection of the sand cap sediment 100 meters to the east and west. The hard sand substrate with considerable shell hash accumulation has appeared extremely stable with no signs of Stamford material protrusion through the cap layer. Elevation stakes and penetrometer tests have produced consistent readings through the fifteen month post-disposal monitoring period. Evidence of gradual intrusion of natural silt surface material over the original sand cap apron region (20–30 meter band) has occurred.

Corymorpha Pendula

In-situ observations during winter and spring seasons have revealed recurrent (1979–80) predominance of the stalked hydroid *Corymorpha pendula,* in epibenthic communities at both New Haven and New London disposal site regions. These organisms may play an important role in boundary layer dynamics and have served as "indicator species" for limits of spoil coverage. A more detailed study of *Corymorpha* distribution and ecology has proceeded as a basis for understanding the recolonization process in relation to spoil impact effects.

During diver transect surveys at both New Haven and New London sites, *Corymorpha* densities and anatomical dimensions were obtained. Behavioral observations focused on the sediment/animal feeding and sessile securing mechanisms. Benthic boundary layer current speeds were also determined.

Densities were determined visually using a .25 m^2 PVC pipe quadrant or by demarkation of a .25 m² area and counting individuals within the boundary. Areas were selected randomly (i.e., swim five seconds and place quadrant on the bottom). Hydroid dimensions (stalk height, hydranth diameter, holdfast depth, holdfast diameter) were measured *in-situ* with a metric ruler. Preliminary data indicate *Corymorpha* is present, as a sessile polyp from February to late May. The remainder of the year it exists in a medusa stage.

Average densities of between 9.0 and 24.0 per .25 m² were observed at the New Haven site and 0.67 per .25 m² from one survey at the New London site. *Corymorpha* has consistently been underrepresented in epibenthic net samples and Smith-McIntyre grab samples when compared to in-situ observations. This may be due to the handling during sampling, sorting procedures and fragility of the animal, since the amorphous remains of some individuals are unidentifiable.

In conclusion, this review has attempted to illustrate the different techniques used to detect benthic impact and long term changes occurring at southern New England ocean disposal sites. Certain observations are first account descriptions of dredge spoil dynamics and behavior of associated biota.

CHAPTER 11

Dredging and Spoil Disposal Policy

CHRIS ROOSEVELT
Stamford Marine Center
United States

Dredging and spoil disposal projects are too often considered in technical and scientific terms without sufficient attention given to the important social, economic, and political factors, which help to define both the need and the justification for the project as well as its impacts. Particularly for federal projects, both the original authorizing legislation of the Army Corps of Engineers, and its annual budget appropriations, reflect a Congressional intent to serve the perceived socio-economic needs of communities for water-born commerce and harbor development. In some instances, such needs become after-the-fact justifications for projects that are possibly politically motivated, at least in part, and suffer from a variety of ills ranging from an assumed federal agency mandate to spend money to an imbalanced decision-making process that is responsive to the loudest voiced interests.

Despite the obvious socio-economic roots of dredging and spoil disposal policy, most contemporary projects are approached on an ad hoc, situation-by-situation basis, discussed in largely technical and scientific frameworks, and adopted with inadequate attention to the full range of economic and resource management interests, particularly on a region-wide and long-term basis.

As but one example, a long line of studies, culminating in the New England River Basins Commission (NERBC), Long Island Sound Study of five years ago, recommended port consolidation and pipe line transportation, particularly for petroleum products, as a means of promoting cost and energy efficiency and reducing the need for maintenance dredging. Yet to my knowledge, not one project proposal has ever mentioned this perspective, and I believe it is also completely neglected in the soon to be issued composite EIS for dredged spoil disposal in the Long Island Sound area. The lack of such planning and the failure to integrate it into the proposal and evaluation process, speaks eloquently of a vacuum in education and political awareness.

Regarding the Long Island Sound study, most of its recommendations would still be sitting on the shelf, were it not for the Coastal Zone Management Act, the Nassau/Suffolk Regional Planning Agency, and the Connecticut Coastal Area Management Program. The Long Island Sound study's biggest failure was in not establishing mechanisms to enhance the public's understanding of the benefits to be achieved through long-range planning for resource management.

The current state of dredging and spoil disposal policy and practice fares little better. What is needed is better region-wide long-term planning, concurrent with an effective political and educational process designed to make the public and decision makers alike, aware of the benefits to be achieved through a balanced evaluation of all applicable social, economic and environmental factors.

Let's look for a moment to past practice in dredging and spoil disposal policy. Much like the Corps' experience with beach erosion, the history of dredging and spoil disposal is replete with examples of man trying to work against, rather than in concert with nature.

In the Long Island Sound area, north shore harbors in Westchester County and Connecticut have naturally been shallow, responding to both net current drift in the sound, and siltation from rivers, run-off and discharges. Traditionally, the historical federal channels in these harbors were established to serve the needs of intense water-born commerce and transportation which were the lifelines of these communities long before the Northeast rail corridor and its successor, Interstate 95.

Environmental impacts were little thought of, except perhaps by oystermen and Indians, and the by-products of the industrial revolution, had only begun to take their polluting toll. Dredging projects were routinely undertaken for both federal and private purposes, and dumping grounds were scattered in many areas around the sound, to efficiently and conveniently serve the nearby harbors being dredged.

Cost benefit analyses, or should we say justifications, were relatively simple; the nearby open water disposal site was cheap, and the economic benefits to industrial and commercial development, jobs, the economy, and transportation were demonstrated.

As the time has passed, the community's focus on the waterfront has been redirected inland to faster and perhaps more convenient methods of transportation such as rails and highways. Water-born commerce has been relegated to heavy construction materials, scrap metals and petroleum products. Concurrently, some industries which needed the waterfront for cooling water, waste disposal or water transportation, have moved inland or to other areas of the country in search of low cost labor, leaving behind outmoded and often abandoned facilities as challenges to adaptive re-use, or obstacles to public access. More recently, recreational boating facilities have grown on the waterfront, and in turn, have been threatened by poorly focused tax laws, and a higher use return potential of office buildings and condominiums. In the process, exceedingly high percentages of our wetlands and shallows areas, have been intentionally or accidently destroyed with direct or related impacts on shellfish beds, fisheries productivity and shoreline protection.

Yet, until very recently, the same cost benefit analyses used years ago, were still used to justify projects to maintain the same federal channels, but not the same users, transportation patterns, or volumes of materials. Without recognition, communities' reliance upon water-born commerce and transportation has changed dramatically, and yet the socio-economic bases have remained the same, perhaps for political reasons.

With the knowledge we now have of the toxic or polluting characteristics of some harbor-bottom sediments, and of their short to intermediate term impacts on marine life when disposed of in open water, decisions to dump some harbor sediment in open water are decidedly examples of intentional pollution. Such decisions today are often political in nature, responding to pressure from limited interest groups, who do not necessarily represent the public's interest in resource management, much less the changing economic mix in harbor useage in our coastal towns and cities.

In a few other instances, more realistic appraisals of the balance between economic need and resource utilization, occur in a framework, which seeks to minimize the intentional pollution.

I would choose to believe and sincerely hope, that the latter process is coming to the forefront in more and more projects, and may soon encompass the kind of long-term region-wide planning referred to earlier.

The Stamford-New Haven project is an example of the political impact of decision making. Proposed in the early 1970s as a decidedly necessary federal channel maintenance project, the east branch of Stamford Harbor, has probably seen the back burner, more times than any other needed project in the Corps of Engineers history.

According to Corps records, the east branch was last dredged to its authorized depth in 1943. Over the intervening years, natural and unnatural accretion, reduced its authorized depth by some 30% to 40%, impeding deep draft navigation, primarily tugs and barges at anything below half tide. Effluent from Stamford's primary level sewage treatment plant, and various industrial dischargers along the east branch, raised the pollution level of its bottom sediments, to among the highest of all Long Island Sound harbors. Because of this heavy pollution load, there was little biological productivity in the east branch, and immediately adjacent harbor areas, and therefore, little question as to whether the dredging itself, would do anything other than enhance the environment.

The questions about this project, related almost solely to what disposal option, if any, was suitable for Stamford's material. Aside from the documented shoaling of the east branch, the need for the project was primarily based on three regular commercial users, offering a total of 110 jobs with a combined annual payroll of over $1,600,000. Two of these users, a bituminous products plant and a fuel and masons material company, had strong business relationships with Stamford area construction firms who purchased their manufactured goods.

Perhaps it should be noted that of the marine owners on the east branch supplying some 600 slips to recreational boating, not one of them spoke of a dredging crisis in the east branch channel.

Some other brief figures that have been recently supplied by a publication issued by the State University of New York at Stony Brook, relate to recreational boating, and this is for Long Island Sound as a whole. But, as they point out, they do show a large emphasis in recreational boating at the western end of the sound, with an estimated 80,000 recreational craft in Long Island Sound, with an estimated value of $368,000,000. This is in 1974 dollars, and the data is from the New England River Basins Commission. In view of inflation, I would suspect that the figure is probably double that today.

Another piece of data from the same study indicates 1977 commercial fish landings from Long Island Sound of some $5,900,000.

I think it should be pointed out, that neither of those statistics cover what I call the "ordinary person, mom and pop" type of operation, as well as recreational use of Long Island Sound. Dollar volume figures are very hard to obtain with regard to the operation of bait and tackle shops, the operation of small boat livery or rental programs, and the operation of party boats.

One thing I should add parenthetically at this point, is that the impact of pollution of the sound on some of these economic sectors, is perhaps significantly greater than many of us realize.

On June 7, 1980 at Stony Brook, the Oceanic Society sponsored a conference called, "The State of the Sound Conference." During the course of an informal question and answer period, Jack Foehrenback, at Stony Brook, related some of his studies collecting 1,000 blue fish and striped bass in the Long Island Sound area, having tested 300 of them and coming up with some very high counts of PCB in the filets of these fish.

As usual, the press immediately hit the panic button, broadcasted it all over the wire services and on all the regional press, and the headlines the next day and really for the next week, talked about PCBs in the most popular sportsfisheries operation in Long Island Sound.

It didn't take long for the word to get back to me via the Connecticut Sea Grant office, which said that not only were they getting a lot of questions on the subject, but that they had documented cancellations of a fairly significant amount of sports fishing and party boat reservations as a result of people's fear of PCBs in the sound. That's just one small example of what scare tactics can do.

It turns out, and I'm speaking secondhand at this point, that Jack Foehrenbach's actual samples, in fact, only found high levels of PCBs in a few of the 300 samples tested and that the rest were at or near the F.D.A. five parts per million level or below it.

In addition, the Stamford dredging project could clearly be justified on at least two environmental grounds: first, virtually every passage of the east branch by tug boats, even at high tide, caused substantial resuspension of the fine grained polluted sediments, which depending on the tide cycle, were potentially distributed throughout the harbor and adjacent Long Island Sound. Obviously, this was no longer a question of leaving polluted materials dormant at the bottom of a harbor.

Second, Stamford's new secondary level sewage treatment plant, had finally come on-line and was boasting of a clean effluent, albeit chlorinated, that was going into a hopelessly polluted east branch.

Throughout the history of the Stamford proposal, disposal was the major problem with the options being reduced year by year. Initially, Stamford's sediments were scheduled to be part of a multiple harbor project, including Mamaroneck Harbor, to be dumped as part of a so-called experiment at the historical Eaton's Neck site.

Following on the heels of the New London case, this Eaton's Neck proposal was decried by Long Island politicians and representatives of the environmental community as a thinly veiled attempt to avoid the E.I.S. requirements of the National Environmental Policy Act.

Quietly the proposal was withdrawn by the Corps, which then settled into a posture somewhat to the effect that if sufficient community support could not be generated for a particular dredging project, then there was little reason to move forward with it in a timely fashion. Clearly, the time for politics was at hand.

In Stamford, this took the form of a revitalized waterfront industry group, who quietly hired an effective and talented gun slinger who undertook a campaign of pressure with the municipal government, with the state D.E.P., with the Corps, and with the Congressional delegation.

In sum, the Industry Association sought to document the critical economic need for dredging. They belittled community and scientific concerns about environmental impact, and strongly advocated open water dumping as a quick solution, saying that the economic benefits outweighed the minimal environmental costs.

The Corps, again the reluctant bride, balked at the division of support and dragged its feet. Largely at the initiative of community groups and the chairman of the Stamford Environmental Protection Board, an attempt was made with the industry group to constructively and imaginatively break the stalemate.

After several informal meetings, ultimately a philosophy of Stamford solving its own problems was adopted and a containment site in the east branch adjacent to an existing city-owned park was proposed.

Initial cost estimates for the containment dike approached half a million dollars, slightly over the cost of the project itself.

Two years in a row, Congressman Stewart McKinney managed to get the necessary appropriation in the public works bill, only to have both bills vetoed by President Carter. The Industry Association, along with many others, lost their patience.

Despite their protests that they did not wish to pollute Long Island Sound, the Industry Association continued to advocate open water disposal, and minimized the risk of impact on the marine environment. This, despite the fact that the Stamford spoils exceeded the applicable E.P.A. criteria for ocean dumping, that little marine life existed in the area from which the spoils were to be dredged, and that preliminary bioassay tests showed high mortality among organisms exposed to the spoils.

It should be noted that the ocean dumping criteria do not yet apply to Long Island Sound and that this is not specifically an ocean dumping example. A recent bill that passed the House of Representatives sponsored by one of the main authors of the political process with regard to dredging and dumping in Long Island Sound, Congressman Ambro, was designed to apply the ocean dumping criteria to Long Island Sound. From what was discussed at the June conference at Stony Brook, it will really not effect the major projects that are scheduled for the sound in terms of limiting the kind of material that will go into the sound. What it will do is impose incredible financial burdens on the smaller dredger, to comply with the various tests that are required in order to meet the ocean dumping criteria.

At this time, the political process really gained momentum with almost daily articles and editorial pieces in the local press citing the waterfront industry crisis and leveling charges against community and environmental spokespersons as "Ivy league pinstripers, wishing to return all of Long Island Sound to its original pristine natural condition."

During this process, the community groups, the Congressional delegation, the state DEP, and the Corps, kept a low profile and continued to search for an environmentally acceptable disposal solution.

Sometime in late June of 1978, this effort resulted in a Corps proposal to conduct an experimental capping project at the historic New

Haven dumping grounds, with the Stamford spoils as the core, material, and the relatively cleaner New Haven Harbor spoils as the cap in a closely monitored, high accuracy, point dumping operation. Virtually everyone jumped at this potential solution.

However, the Corps had no experience with capping, although the New London operation achieved a layering effect through careful phasing of segments of the Thames River. Additionally, a capping project in a Norwegian fjord was cited as a successful example. Several hard questions were asked regarding the proposal. Did we know enough about the physical characteristics of the Stamford and New Haven spoils to assure that they would not interact like a stone falling on a mound of jello? Shouldn't we experiment first with relatively clean spoils before risking marine life with Stamford's highly polluted spoils?

After several informal and public meetings designed to resolve these questions, and as most will now agree did not, the Corps issued in July 1978, its formal notice of the combined Stamford-New Haven project.

Continued attention to these questions and to the design of a detailed monitoring program, occupied much of the remainder of the summer and fall of 1978. Aside from still not having a firm grip on the physical characteristics of the proposed constituents to the project, other complications arose.

Several members of the independent scientific team brought together to design the monitoring program, balked at the lack of freedom to do just that. Once they were agreeably brought back into the process, other questions arose as to the total volume of the project.

Originally, it was slated to involve 65,000 cubic yards from Stamford and 115,000 cubic yards from New Haven, a close 2:1 ratio that seemed adequate for capping. Then during the late fall of 1978 it was acknowledged that to bring Stamford to its authorized depth would involve 105,000 cubic yards, reducing to a very questionable minimum, the capacity of the New Haven spoils to cap the Stamford spoils. A compromise was worked out limiting the Stamford dredging to approximately 76,000 cubic yards, and designing a two pile disposal operation, wherein one would be capped with fine grained New Haven materials, and the other with outer harbor courser grained sand.

Beginning in March 1979, the project got under way with an appropriate spring and summer window for the oyster spawning season. The bulk of the Stamford project was completed in October 1979, with New Haven closely following.

From the other chapters in this unit, I believe you have learned the rest, including the damage Hurricane David did to the fine grained clumpy cap. Unfortunately, you did not learn that because much of this project was carried on during the fall months, there was inclement weather incurred on many days. The barge traveling from Stamford all the way up to the New Haven site was periodically beset with heavy seas, a tremendous amount of wash-out of the barge occurred, and in fact, there is at least one documented short dump that occurred off the Norwalk Islands.

Suffice it to say that the Stamford-New Haven project has taught us something about capping procedure at that depth and at that location in the Sound.

Despite the dominance of the political process in this project, social and economic benefits were achieved and our scientific and technical knowledge has been confirmed, if not enhanced.

Some summary comments are in order. First, if only because of the wide-spread acceptance of the oyster spawning window, shell and fin-fish and crustacean considerations are finally rising to an acceptable level of importance commensurate with their actual, if not potential, fishery values. Unfortunately, such concern has yet to spread to rec-reational fishing and boating and the important economic contribution to the region that they provide.

Second, there has been little effort to date to compare the relative economic contributions of the marine recreation business with water-front commerce and industry and the comparative growth and decline that have occurred in each respectively over the past twenty to thirty years.

Third, it is still evident that little effort is being made to apply long-term region-wide planning processes, such as transportation or port consolidation plans to any shoreline community, much less to specific dredging projects.

Fourth and finally, it is evident from the Stamford-New Haven project that many dredge-spoil disposal decisions are largely political and economic in nature, not just from their inception, but also in the final solution of serious technical, scientific, and environmental ques-tions.

While the democratic process is unquestionably this country's great-est strength, it works best in balance with informed participants con-tributing to the rational solution of issues. Therefore, what lies ahead of us is a rededication to public and decision-maker information, edu-cation, and awareness.

CHAPTER 12

Commentary
Waste Disposal Policies:
How Effective Are They?

VYTO ANDRELIUNAS
U.S. Army Corps of Engineers
United States

As a member of the Army Corps of Engineers my credentials here are those of a bureaucrat. My job is one of trying to find a public policy that will satisfy the requirements that are laid down for us, which are to make sure that no one comes to grief in our navigable channels and also to insure that there is not an unacceptable level of degradation in the marine environment due to our activities.

I am sure, as you all know, that most of the rules and regulations and laws are drafted by folks who are very clever with the language they use. Dr. Saila, I think, set the table very well (see Introduction to Unit 3) by asking whether or not the type of activity we are discussing really fits into the context of pollution in the marine environment.

I can assure you that whether or not we can ascribe the gravity to disposal operations that we do to other activities I can not be guided by the good words of scientists alone in deciding whether or not I have to be concerned with those things. I have to be concerned a long as the media says that we have a problem.

So in reaching the public policy decisions it is not for us to be concerned with simple scientific facts alone but to be concerned also with the perspectives that come out of the public arena, and those are shaped and molded in many ways. I think Mr. Roosevelt (see Chapter 11) has given you a very good account of a relatively small project involving an estuary where industrial wastes and domestic wastes have been deposited over a period of at least half a century.

I have to accuse Dave Shonting (see Chapter 9) of unwittingly falling into the trap of using words such as "highly toxic" and "very great problems" all of which I think all of us from time to time slip into in discussing these types of things.

The fact that heavy metals are toxic in certain concentrations is a given. But whether or not heavy metals, when associated with sediments that are deposited in the marine environment, are also toxic is still debatable. I think we are still lacking in our approach, in our methodology, of evaluating those impacts. For example, most recently we have been using bioassay techniques. There are regulators who are trying to judge whether or not there is an unacceptable impact on the ocean outside of the prescribed dump site by placing both the ellutriates and the sediments themselves in an aquarium with selected species. No one yet, has been able to figure out how to interpret results that come from such tests.

So we get back to the provocative questions that I would like to ask. Are we really acting in the public interest in the way we are carrying out public policy today, or are there things that we should be doing better, and how can we do those things better? How can we get a better understanding from our politicians? How can we get the Office of Management and Budget to understand what it is we are trying to do?

The Corps of Engineers, as everybody knows, is not an oceanographic institution. The Corps of Engineers probably does not belong in that arena, but we have been thrust into that arena and we were given $30,000,000 to conduct a five-year research program. It takes two years to get started on any decent research program at least. By the time we found out what the questions were that we needed to address, the five years had essentially slipped by.

We did develop a tremendous amount of information that everybody needed. I have a feeling that that information is going to be much more useful to people who are dealing or have yet to deal with the impact of solid waste on terrestrial areas. Nevertheless, we made a start. The attitude of the Congressional Committee, however, was, "We gave you $30,000,000, why do you still have a problem?"

We need to continue to address certain questions. I ask you, who come from different parts of this country and from other parts of the world, to look at it from a policymaker's point of view if you will, and given all these constraints and the world that we live in, are we truly muddling through in the public interest, or how could we do it better?

Discussion

Saila: I think no scientist ever feels that he has enough information about anything that he likes to do. On the other hand, I do personally believe, that the role of a scientist — an applied scientist — is sometimes to provide advice to the community of decision-makers. I would like to open this discussion with my personal assertion that at present we have adequate scientific information for a rational assessment to society of the consequences of at least some types of marine dredge spoil disposal. I think there are elements which are missing in the social, political and economic arenas from which decisions need to be made.

Comment: I would like to know the cost of dredging and the cost of research. Do you happen to know what it costs to do that dredging and what it costs to do the research on that?

Andreliunas: The cost of dredging is usually reflected in market prices. The cost of dredging has not increased appreciably over the years, in fact, it has probably fallen behind most price indices for the construction industry. There is, however, an additional cost of hauling involved in disposal and perhaps for environmental reasons that could be looked at as an additional cost imposed by managers. It is very difficult, unless you take a specific job and give specific questions, to come up with an exact added cost, but there is an added cost for hauling, which can be computed on the basis of the additional time, fuel, and manpower involved in that activity.

Regarding research, the cost of an annual monitoring program which covers the entire coast of New England, such as we are now doing, and involves two cruises and other special features has been approximately $600,000. Considering what vessel time costs, that is not a great deal of money. The monitoring, all the added work, associated with the Stamford-New Haven project, was probably in the neighborhood of twenty-five to fifty percent of the actual cost of doing the work, depending on how the costs were separated.

Roosevelt: What was the total Stamford-New Haven cost then?

Andreliunas: Somewhere around a million and a half dollars.

Roosevelt: My recollection is that at least the first Thames River project, was around $1.80 a cubic yard, and the second one was a little bit higher because the contractor had to bring his equipment back into the estuary about two or three months after the first project had been completed to take another four feet out.

Comment: We spent about $8,000,000 to dredge the Thames River and, as I understand it, about $30,000,000 to do the research on the two projects together, if that was correct.

Saila: No, that was a nationwide program. The Dredge Material Research Program (DMRP) was a nationwide waterways experiment station sponsored program, costing $35,000,000 spread over five years. It involved terrestrial sites and marsh creation, and a lot of elements.

Roosevelt: It was geographically focused, I believe, in the Gulf Coast areas and in the Great Lakes.

Saila: Primarily, yes.

Comment: I would like to expand a little bit on what Dr. Saila said about there probably not being a lack of scientific knowledge here. Ed Goldberg talked yesterday about contaminants with similar capacities, end points, and generic sites and said there is no way we could get end points for the 50,000 things you could think of. As I understand it, when you deal with dredge spoils, the basic problems are things like PCBs and heavy metals, and for those things we do have end points. Granted, that it's not a simple function of metal concentration, because it is a function of speciation.

The first approach seems to be to determine allowable levels in, for example, fish or shellfish. I have two questions. First, are there experiments in progress to determine what effects dredge spoil dumping have on shellfish or fish that would be near dredge materials. Second, if you had that information, would it do any good or would it just be argued forever between the environmentalists and the dredge slingers as to whether their data was worthwhile?

Stewart: The first question, what are the limits of heavy metal contamination that are allowable at a dredge site? Those are being approached after about three years of data analysis, as part of the mussel watch program. It is very perplexing not only because of species selectivity and different tendencies of one species to select for a certain metal more heavily than another but also because you can not determine a definite effect of the heavy metal loading at any particular time and divorce it from all the other environmental fluctuations — whether it be severe run-off in high times of precipitation in the spring or something related to dredging effects. So you have to be very careful in choosing limits.

Another condition you must recognize, is that when the dredge spoil material is in the harbor, it is in a shallow water environment, a high energy zone. It is prone to storm disturbance routinely, it is disturbed by prop wash of vessels, and it is closest to any of the spawning grounds; the estuarine regions, which are focal points for migration of many species and are overwintering areas.

Prudent disposal methods could be to remove material and put it in a biologically less important region of the ocean and contain it. It's a philosophical and subjective point, but everybody avoids reference to the process as cleansing the estuary. In effect, maintenance dredging may be doing that. You may be increasing flushing rates, and reducing containment loads in areas of greater human access.

Comment: Have values of PCBs in mussels exceeded acceptable limits for human consumption at any time at dredge spoil sites?

Saila: Dr. Goldberg, how great is the probability of mussels hanging two meters above a dredged material site, detecting or picking up any perceptible amount of these trace metals or organics?

Goldberg: You have asked me a very general question about a whole group of metals. It is a very difficult question to answer. Let me turn it around and give you an answer along these lines. A group was concerned about cadmium in dredge spoil materials. Knowing the cadmium in the dredge spoil materials, the leakage to the water, and the accumulation by oysters, they made a model and concluded that eating oysters taken from this dredge spoil area constituted a potential danger.

Clearly an important step is to monitor what the levels are of this heavy metal in commercially consumed fish or shellfish. To get back to your original question, if you suspend mussels over the dredge site in which heavy metal leakage is expected, I think in some cases the mussels will pick it up. This is intuitive; it is not based on any knowledge that I have.

Comment: I have been very impressed with the voluminous documents that have come out of the $35,000,000 dredge spoil research program in the Corps of Engineers. Five years seems like a fairly long time, but, in fact, two years of that is spent in developing programs and identifying problems.

I would like to ask the provocateur the question, first of all, does this sort of time frame allow for any research on new problems that come along the way.

My second question which is directed to the panel in general concerns the bioavailability of some of the metals or, for that matter, chlorinated hydrocarbons. I am thinking of those substances that are in annex 1 of the London dumping convention. In our legislation we have actually had to ban the dumping of certain of these metals, such as mercury and cadmium and PCBs as well as DDTs, if they exceed a certain limit. We have set regulatory limits on these, but we are beginning to see that this is really not too meaningful in the biological sense, because normally all of the material that is present in the concentrations that you find is a total amount of cadmium or a total amount of mercury, and is actually biologically unavailable — it is fixed. On the other hand, it may be available; we have seen this happen.

Andreliunas: You are asking if any of the problems that have come out of our research endeavors are being channeled into some other research program? My feeling is that they would be if such research programs were supported. We are, whether we like it or not, continually driven by crises. In the New York Harbor Crisis, or whatever you want to call it, which arose recently and was highlighted through Senate hearings, the data on PCBs, particularly in the New York area, was poor. I think we all realize that there was certainly deficiencies in that area. We do not know an awful lot about PCBs and their effect on humans generally. That certainly would be exploited if, in fact, the Congress authorizes additional research.

Such research is now being directed towards the EPA, as opposed perhaps to NOAA which has a responsibility for research under the Ocean Dumping Act. We have never been funded to do that particular thing. It is strange that although the Ocean Dumping Act addresses itself to dredge material and talks about research, the only authorized funding for such research has been to the Corps of Engineers, who are regulators.

Stewart: I might add a comment. I did not present data on heavy metals, Dr. Feng conducts the study. However, some interesting aspects of the research are understanding what is occurring in heavy metal concentration within particular shellfish species. You may get a spike of one particular metal in a species that may not be coincident with any pollutant activity or run-off, it may be entirely a metabolic phenomena. That's obscure in some respects but it occurs. Also, there are very definite signs of depuration over a period of time. So I'd go back to what Dr. Goldberg said, if you want to know the source and the level of danger, go to the heart of the estuary and determine what some of the prime commercial stock source levels are, and then by dilution in a dredge disposal operation, you might be able to draw some connections.

Comment: I've got two questions. In terms of the very large scale, you have an experiment going on in New York with a million cubic yards of so-called "hot material" to be capped. Was any work done on initial release or uptake for the species, such as fish and lobster, that are attracted to the site? Until the cap is in place, you essentially have open water disposal. It takes a period of months between the time of discharge and capping. Is capping that effective a management tool?

Shonting: It is clear that a capping operation has a critical time period in terms of how much you have to move and how far you have to move it. However, I am not familiar with this large-scale operation. If you can make some judgments about how long it takes for the undesirable material to leak into the ocean, then you can probably conjure up a maximum-size capping operation that you'd wish to undertake. So I think that one of the things that the management aspect of a capping undertaking involves is that you have to think of the time that it takes to do it. You have to put the stuff on the bottom first and then cap. You cannot build it like blocks, piecemeal, although perhaps there are ways of doing this.

Andreliunas: There is not yet an operation going on in New York, but it is being planned. There are several strategies being discussed. I think that borrow pits were mentioned as a good possibility. New York is trying to benefit from our experience in Long Island Sound for conducting a larger scale capping operation. What you say is true; the material is considered to be more contaminated there, and there is an opportunity for colonization which may have impacts associated with it.

I want to point out that the aborted experiment which was to have been done under the Dredge Material Research Program (DMRP) in Long Island Sound involved putting down a fairly large amount of Stamford material, which was considered to be contaminated, and observing the impacts associated with leaving the material uncovered. This is probably a failing of our New Haven project; we covered all of that material and did not leave a discrete portion of material which was considered to be unacceptable so that we could, in fact, observe what happened to it. I think that is our piece of information that we are still lacking and which we need.

Roosevelt: May I suggest a quick look at the historic Stamford site which is just off the Stamford breakwater about one and one-half miles. While it is not thoroughly documented as to what was put there, it has been estimated that the site has material that is at least as bad if not much worse than what was recently taken out of the east branch of Stamford Harbor. I would think that a twenty year experience of seeing that lie in one place would be a very good way of testing exactly what happens on a long-term basis.

Andreliunas: This is a good point. The DMRP, over one season did look at the historic Eaton's Neck dumping ground, which up to that time, according to our records, had received over 20,000,000 yards from the New York area. We can assume that that material showed similar characteristics. We have found — and I hesitate to cite numbers — that the lobsters on the dumping sites show no more contamination than lobsters far away from the dumping grounds. In fact, the levels of heavy metal, or whatever you are measuring in the lobster, tend to go along with the age; I think this correlates to what has been found with other long-life species.

Incidentally, the dumping ground at Eaton's Neck coincides with the most heavily lobstered areas in Long Island Sound. In fact, all of our dumping grounds coincide with all of the heavily lobstered areas. There is little publicity associated with that, and I think perhaps the less the better at this point, because we have already been through the swordfish business and the cranberry business, and we do not need a lobster thing.

Comment: Just a quick question to the gentleman from the Corps of Engineers, maybe more or less on a different track sir, as I want to satisfy a question in my own mind.

I understand the state of Massachusetts is considering depositing thousands of tires offshore for a tire reef, supposedly to enhance the fisheries. It might be a substitute way of getting rid of thousands and thousands of tires and thus would be almost an indirect method of intentional dumping. I am wondering whether there has been any research done on what would happen with thousands and thousands of tires that would deteriorate in time. Would chemicals be released from them? Has there been any study whatsoever about this type of indirect intentional dumping?

Andreliunas: We are currently sitting in judgment on that one. There is an application for a permit to do just that and there is a considerable amount of research on the effect of tire reefs. I think that the issue at Gloucester is the concern of fishermen for the effect on their trawling gear rather than the release of toxins.

National Marine Fisheries Service is the repository of much information on tire reefs. Quite frankly, the contention is not whether to put tires down there, but where is the best place to put them if they are needed at all. A decision will be made in months, I am sure.

Comment: In light of what has been said about capping operations and erosional processes affecting the cap material, I am not convinced that we can be sure that capping does indeed work, for in the case of erosion we have been dealing only with a tropical storm, which is of less intensity than a hurricane. Can we be sure that that material will remain as a cap or will an event, such as a tropical storm or hurricane, perhaps remove the capping material and the so-called "hot material" at the surface?

Shonting: A storm passing over a site can cause a large transfer of wind stress and large waves, but the actual magnitude of the waves is generally a function of the fetch and the duration of the winds. For example, if you had a storm passing over Long Island Sound and the winds were directed north-south or perpendicular, to the axis of Long Island Sound, you would get relatively small waves. If the waves were parallel, you might get very large waves and very high energy on the mound itself. These factors have to be considered.

Comment: With monitoring programs going on, you would know if there are any changes in the cap which would require putting on additional clean material. That is one of the primary objectives of the New York job.

Another thing is that the previously mentioned storm occurred when the spoil mound was fresh. Dr. Stewart has talked about the bioturbation and the flattening out of the mound over a period of time. It is our feeling that it really was the roughest elements of the mound that caused the erosion. The north mound of sand to the surface was actually deeper than the south mound. Over a period of time that mound would probably consolidate anyway, and probably would not be eroded the way it did when that storm occurred.

UNIT FOUR

Chronic Pollution: A Case Study of the Southern California Bight

Photo courtesy of Willard Bascom

OFTENTIMES WE FOCUS OUR ATTENTION, OUR RESEARCH DOLLARS, AND MUCH of our work on oil spills, dredge spoil disposal, or on toxic materials that are put into our nation's waterways, and rarely do we recognize that the largest input of pollutants into our coastal waters is ordinary sewage. Needless to say, the volume of sewage produced and put into our coastal waters is immense. One must recognize that the total number of people living in our largest cities in the coastal zone of the United States utilize marine or estuarine waters in one fashion or another for the final resting place of their sewage.

Generally speaking, we do not realize that sewage is much more than domestic or home wastes which are enriched in carbon, phosphorous, and nitrogen. Many of our cities and towns also have industries that dump their waste directly into the sewage systems. In some cases it is a fairly major and toxic input, as in the case of Los Angeles where there is a DDT manufacturing plant that over a period of years has put DDT directly into the sewer system. Here in Rhode Island there are scores of small metal plating firms that dump their wastes directly into our sewage treatment systems, thence into Narragansett Bay. This sewage is our most ubiquitous effluent, and it is a highly complex, variable pollutant.

Before 1972 the only law governing the discharge of wastes into coastal waters was the Harbors and Refuse Act of 1889. That legislation was refined through the years by the courts but lacked a central philosophy on how one should control pollution in our nation's waterways. In 1972 Congress passed the Water Pollution Control Act, a bill that attempted to codify national waste-cleanup goals, set standards for pollution control technology (i.e., secondary sewage treatment for all marine outfalls), and gave deadlines for implementing these technologies. Some municipalities balked at these rigorous standards, and in 1977 Congress amended the Water Pollution Control Act to allow municipalities to obtain a waiver from certain control technology if they could prove that their present disposal practices did not greatly impact the utility and the biological populations of the receiving waters. This waiver of secondary treatment can be given by the Environmental Protection Agency (EPA) if a series of eight caveats are met. Among them is one dealing with public health and another dealing with the meeting of state water quality standards. Another one, which may make some ecologists cringe, concerns obtaining a waiver if the sewage treatment operation does not affect "the balanced and indigenous populations of the area." Initially, approximately 100 to 150 applications for waivers were submitted to the EPA, and they have now been reduced to about 70 or 80 valid permits. The EPA has prioritized those communities, towns, and cities in going through this waiver process. Representatives of the National Oceanographic and Atmospheric Administration and the U.S. Fish and Wildlife Service are aiding in the waiver evaluation process.

There is a rather interesting paradox concerning the 1977 amendments and the section 301(h) waivers. Many of the larger municipalities and cities, e.g., Los Angeles, had the data from which to go through this waiver process, but many of the smaller cities and towns, e.g., Narragansett, Rhode Island, did not go through the process because they did not have the data and they were not funded well enough for research and science to apply for the waiver. Thus, large cities and towns, which may have the largest impact, can apply for these waivers because they will have the data to do it, while the smaller towns and cities may not apply for waivers even though they may be the ones most applicable to the process.

Finally, we should keep in mind, and this is the Achilles heel of the Environmental Protection Agency, that the EPA has the same problem that the Atomic Energy Commission had several years ago. The EPA is not only the licensor, but is the promoter of sewage treatment plants. They fund the construction, and they process the grants and applications, but at the same time, they are the people that are regulating the industry itself. So much for checks and balances. Hopefully the logical tug-of-war by the authors of this section will point to a better balance.

Eric Schneider
Center for Ocean Management Studies
United States

CHAPTER 13

The Effects of Sludge Disposal in Santa Monica Bay

WILLARD BASCOM
Southern California Coastal Water Research Project
United States

This report contains a summary of scientific findings that relate to the effects of discharging sludge from the Hyperion Wastewater Treatment Plant via the "seven-mile pipe" into Santa Monica Bay. Most of the data used here were obtained in 1978 and 1979, but the situation does not appear to have changed much during the previous five years and thus probably will not change greatly in the next few years regardless of what action is taken. This paper is a shortened version of one which appeared in the 1980 "Annual Report of the Southern California Coastal Water Research Project."

Nothing in this report should be construed as an argument for or against the continuation of sludge discharge. Hopefully the existence of well-documented measurements of the present physical, chemical, and biological situation will contribute to a rational decision. At least it will make it possible to compare the ecological advantages and disadvantages of the present site with those of alternative sludge disposal sites or methods.

SANTA MONICA BAY

Geography

Santa Monica Bay, immediately west of Los Angeles, extends over 40 kilometers (27 miles) between headlands; its total area is about 500 square kilometers or 230 square miles. The sea floor of much of the Bay is flat and muddy, but there are several areas of low-relief rocky outcrops. Two large submarine canyons are the most prominent hydrographic feature of the bay. The shoreline is composed mainly of sandy beaches, which are much used for recreation, especially in the summer when waves are low and water temperatures are high. The tide range is from four to seven feet so that tidal currents are superimposed on the general coastal currents.

Several cities, including Santa Monica, Malibu, Marina del Rey, El Segundo, and Redondo Beach, ring the bay. Santa Monica Bay opens into a series of alternating basins (with depths to 2000 meters) and mountain ranges that occasionally break the surface to form the channel islands. The distance from shore to the abrupt continental slope into deep water is about 240 kilometers (150 miles). The very large open embayment between the escarpment and the shore and extending from Point Conception to Cabo Colnet is known as the Southern California Bight. Its area is about 100,000 square kilometers. Thus, Santa Monica Bay is part of this large Bight and receives some protection from north Pacific swell because it faces southwest and is partly in the lee of offshore islands.

In September 1957, after extensive design studies that paid particular attention to the effects of waste discharge on the adjacent water and sea life (by the standards and available information of the late 1950s), the seven-mile sludge outfall was put into operation. This small pipe (two feet in diameter) extends ten kilometers offshore from the Hyperion Treatment Plant in El Segundo to a point near the head of Santa Monica submarine canyon, almost in the center of the bay where the water depth is 100 meters. The intention was to dispose of waste solids well offshore in this steep-sided trench in the sea bottom where they would create only minor problems and have a reasonable chance of being carried further offshore by currents so that organic buildup on the bottom would be minimal.

Redondo Canyon, about fifteen kilometers further south, is a similar but larger hydrographic feature. Because its head comes closer to shore, it traps much of the sand and natural organic material moving along the coast in the Redondo area. The offshore end on this canyon intercepts some of the excess organic material moving northward along the Palos Verdes peninsula from the Los Angeles County outfall.

The water quality in Santa Monica Bay as a whole is influenced by many factors, including two large wastewater outfalls (in addition to the sludge outfall). These are the Hyperion outfall, which discharges five miles offshore — only about two miles from the sludge discharge point, and the Los Angeles County outfall, which discharges about eight miles southeast of Point Palos Verdes. Some of the latter's (greatly diluted) wastes are carried by currents into the southern end of the Bay. The bay is also affected by discharges from power plants and an oil refinery as well as land washoff by rain storms, harbor discharges, aerial fallout, and the effects of many people using the beaches and small boats.

The position of the sludge particles on the bottom is largely controlled by the topography of the bottom and by the currents that flow in and above the canyon. We found it necessary to rechart the area using LORAN C for position and a 50-kHZ echo sounder for depth.

Oceanography

The southward-moving California current (the major current off our coast) flows well offshore, but some of its water turns toward the coast opposite the Mexican border to become part of a great counterclockwise eddy that mixes with warm waters moving up from the south. This mixture flows northward through the Bight and along the outer edge of Santa Monica Bay. The residence time of water (the average time that a particle of water stays in Santa Monica Bay) at the depth of the wastefield is believed to be about one month. For the mixed-layer waters of the Southern California Bight as a whole, the residence time is about three months.

The motion of the water over the relatively flat bottom of Santa Monica Bay is less consistent than along the relatively steep offshore slopes of the open coast, which tend to organize water motion. The currents in the bay often change direction and speed; sometimes there is one large gyre (rotational water motion), and sometimes there are two small gyres. When there are strong easterly winds (Santa Anas), the surface waters are blown offshore and replaced by cooler nutrient rich waters from beneath, a condition known as upwelling.

The density of the water at various depths in the bay, which controls the height to which the waste plume from a pipe rises, depends not only on the water temperature but on its salinity. Both of these change with the season. When the water is mixed by winds and waves, it is nearly the same density from the bottom to the surface. When this occurs, the plume rises higher and breaks the surface; under these conditions, the dilution increases to as much as 300 to 1. This condition rarely exists for more than three weeks at a time; it generally occurs in the winter months but not every year. The other extreme is a large difference in density between surface and bottom water; this is a "highly stratified" condition, which may or may not have an abrupt change in density at mid depth. During this condition the plume does not rise very far above the bottom, and there is less dilution (perhaps 100:1). Most of the year ocean density conditions allow the plume to rise to mid depth and the dilution is about 200:1.

The strongest observed subsurface currents observed in two to five week intervals were found over the shelf inshore of the seven-mile outfall, where maximum speeds were about 35 cm/sec, and the median speed was about 10 cm/sec. The weakest currents were observed near the bottom on the shelf and deep (about 385 meters) in the canyon. Closer to the outfall, in 168 meters of water, the near-bottom currents in the canyon reached speeds as high as 45 cm/sec, but the median speed, at 8 cm/sec, was weaker than that on the shelf.

In all areas, the shape of the sea floor appears to play an important role in the direction of movement of the currents. Generally the principal direction of movement is approximately parallel to the local contours. The exception to this occurred in near-bottom waters on the flat shelf, where the primary motion was directed toward the head of the canyon.

Typical variations in current movement are likely to be 10 cm/sec over the shelf and in the shallow part of the canyon; 7 cm/sec in the deep canyon and over the slope; and 3 cm/sec near the bottom on the shelf and on the slope most of the variations occur over periods longer than one day. Near the bottom at the deeper canyon station, the variation is predominantly of tidal periodicity, but at the shallow canyon station it is largely composed of harmonics of the semidiurnal tidal period. These data support the hypothesis that canyons can "trap" internal waves and intensify the near-bottom currents up and down the canyon.

The levels of dissolved oxygen (DO) in the water near the outfall are of interest, because oxygen is available to sea animals only in this form. There has been concern in the past (especially in restricted inland waters) that wastes that require oxygen for chemical or biological reactions (COD or BOD) would use up the oxygen in the water that is needed by sea life. However, in Santa Monica Bay and elsewhere on this coast, the water usually contains far more oxygen than needed to support both the animals and the chemical reactions.

In the canyon bottom where the percentage of volatile solids and the biochemical oxygen demand of the waste sediments is very high (up to 27,000 ppm), we find a 30% decrease in the biomass of the benthic animals. However, the presence of large numbers of fish in this area suggests that there is adequate oxygen in the near-bottom water for most animals in spite of the large BOD of the sediments.

Three previous surveys were made specifically to establish a basis for ecological comparisons. SCCWRP's *60 Meter Control Survey* of the southern California coast from Point Conception to the Mexican border established 29 stations at a depth of 60 meters that are believed to represent original pristine conditions. (Word and Mearns 1978). The following year, a 300-station survey of *Life in the Bottom* was made that extends from Point Dume to Dana Point for depths of 20 to 500 meters (Bascom 1978). More recently a survey of life in submarine canyons (referred to in table 5) has been partially finished and will be reported in a subsequent paper by this author and others.

Ecological conclusions expressed in this paper were reached relative to these other data bases. "Normal" conditions are further defined and the combined data are presented by Bascom, Mearns, and Word (1978).

In all of these, we have relied on the Infaunal Trophic Index devised by Jack Word of our staff as a means of numerically describing the feeding habits of a community of small animals that live in muddy bottoms (Word 1978). This index uses fifty-two indicator species that are divided into four categories based on their feeding preferences. They obtain their food immediately above the bottom, on the bottom, just below the surface and deep in the bottom. The number of individuals in each of these four categories in any sample of mud is determined and the result used in a simple equation. The index number is high where the first group (water-column feeders) is dominant and low when most animals are in the last group (deep in the bottom). The result of this method is that it is possible to obtain consistent results that have a numerical value and can be used to chart variations in bottom conditions.

Material Discharged and Where it Goes

The average daily discharge from the seven-mile sludge line for the last few years in millions of gallons per day (MGD) is as follows:

- 1.2 MGD of screened, digested primary sludge
- 1.0 MGD of waste-activated anaerobically-digested secondary sludge
- <u>2.5</u> MGD of secondary effluent
- 4.7 MGD total

The average solids content over the last eight years has been 8,900 mg/liter (parts per million), somewhat less than 1% by volume of the material discharged. Details of the average yearly concentrations and tonnages of possible pollutants are given in tables 1 and 2. These are taken from our recent annual reports. Table 3 gives the concentrations of EPA Priority Pollutants in the seven-mile discharge for one day in July 1978. This is similar to the data in table 1 except that it is for the volatile or extractable man-made organic materials in a single sample.

About 57,000 metric tons (dry weight) of solids with an approximate (wet) volume of 285,000 cu m is discharged each year.

Upon discharge, the momentum of the effluent flowing from the pipe carries it outward while its relative warmth and low salinity cause

TABLE 13.1

Average Concentration of Constituents Discharged Through the Seven Mile Outfall*

	1971	1972	1973	1974	1975†	1976	1977	1978	1979	Average
Flow (mgd)	5	4.6	4.8	4.7	4.3	4.1	4.6	4.6	4.8	4.6
mg/liter dry weight										
Total solids	7,900	8,600	8,500	8,400	10,300	9,900	4,400	8,400	7,100	8,700
Oil and grease	760	636	922	900	970	697	608	511	400	710
Ammonic nitrogen	160			300		411	349	302	230	290
Total phosphorus	130			663	80	141	189	247	266	240
Cyanide (CN)		0.10	0.11	0.53	0.67	0.44	0.70	0.24	0.18	0.37
Phenols					0.029	0.61	0.58	0.46	0.53	0.41
Silver	0.03	0.29	0.8	0.4	0.8	0.86	1.88	0.69	0.70	0.72
Arsenic		0.03	0.27	0.18	0.29	0.28	0.26	0.24	0.22	0.22
Cadmium	0.23	0.42	0.98	1.28	1.17	1.31	1.30	1.25	0.80	0.96
Chromium	2.1	8.23	18.2	15.1	11.7	11.66	12.8	4.9	5.3	10.0
Copper	12.2	7.58	13.6	13.9	16.8	16.8	15.5	14.7	8.7	13.2
Mercury	0.1	0.125	1.4	0.15	0.108	0.098	0.131	0.059	0.062	0.25
Manganese	1.6	0.67	0.37		0.19					
Nickel	2.6	2.04	3.7	3.1	3.1	3.6	4.1	5.3	3.4	3.4
Lead	0.51	1.8	1.57	1.13	2.05	1.65	2.14	5.99	4.89	2.42
Zinc	16.5	10.74	27.0	23.0	23.1	21.1	28.6	22.8	15.0	21.0
Selenium		0.25	0.45	0.4	0.27	1.16	1.68	0.69	0.05	0.63
µg/l										
DDT			4.03	2.59	6.49	6.78	1.40	2.72	0.9	3.7
PCB			25.4	3.30	15.3	34.8	20.3	31	10.8	20.4

Source: H. Schafer

*As measured by the Hyperion Treatment Plant Laboratory
†Digester cleaning in 1975 significantly but temporarily raised the average suspended solids.

TABLE 13.2

MATERIAL DISCHARGED FROM THE SEVEN-MILE OUTFALL

	NUMBER OF YEARS MEASURED	AVERAGE PER YEAR FOR LAST 8 YEARS	APPROX. TOTAL IN 22 YEARS
		Metric tons dry wt	
Total solids	22	56,700	1,250,000
Oil and grease	22	4,800	106,000
Ammonic nitrogen	5	1,930	42,000
Total phosphorus	6	1,530	34,000
Cyanide (CN)	7	2.5	55
Phenols	4	2.7	59
Silver	7	4.6	101
Arsenic	7	1.4	31
Cadmium	7	6.3	139
Chromium	7	67	1,470
Copper	7	88	1,940
Mercury	7	1.72	38
Manganese	4	4.2	92
Nickel	7	21.0	480
Lead	7	13.4	295
Zinc	7	138	3,040
Selenium	7	4.4	97
Chlorinated hydrocarbons		*Kilograms*	
DDT*	7	25	550
PCB*	7	138	3,040

SOURCE: H. Schafer
*Based on average after 1972

it to rise. As this jet of waste moves out over the canyon and the water deepens, diffusion and mixing with the surrounding water cause it to expand while currents modify the direction of the resulting plume. Dilution by a factor of 50 to 150 probably takes place in the first 200 meters. Note that the seven-mile outfall discharge point is not at the head of the canyon, which is nearly a kilometer farther north, but at a point along the canyon where the walls are much steeper (see fig. 1).

To determine where the discharged material goes, project personnel directly sampled and measured bottom materials over a wide area. We also collected data for use in theoretical estimates by measuring the currents in the area and conducting settling-velocity experiments in the laboratory using effluent-seawater mixtures to determine how rapidly the particulates fall to the bottom. The latter measurements were used in a mathematical model to obtain a computer solution to the question of how much solid material would be deposited on a flat bottom at outfall depth.

TABLE 13.3

CONCENTRATIONS OF EPA PRIORITY POLLUTANTS IN TWENTY-FOUR HOUR COMPOSITES
OF THE SEVEN-MILE EFFLUENT COLLECTED JULY 1978

Flow (liters/yr $\times 10^{11}$)	0.064		*Extractable Organics* (μg/liter)	
pH	7.3*		2-chlorophenol	8 ± 3
General Constituents (mg/liter)			4-nitrophenol	90 ± 80
Total suspended solids	12,000		Phenol	700 ± 60
Oil and grease	610*		Pentachlorophenol	<10
Ammonia nitrogen	280		Bis (2-ethylhexyl) phthalate	13 ± 8
Total (K) nitrogen	740		*Miscellaneous*	
BOD	4,700**		Asbestos (10^6/liter)	11,000 ± 0
COD	8,700**		Cyanide (μ/liter)	300
Fecal coliform (MPN/100 ml)	3.9×10^6		Phenol (mg/liter)	0.24
Volatile Organics (μg/liter)			*Trace Metals* (mg/liter)	
Benzene	<10		Antimony	0.23
1,1-dichloroethane	<10		Arsenic	0.26
1,1,1-trichlorethane	<10		Beryllium	0.002
1,1,2-trichlorethane	<10		Cadmium	1.3*
Chloroform	<10		Chromium	12.8*
1,1-dichloroethylene	<10		Copper	15.5*
1,2-trans-dichloroethylene	145 ± 5		Mercury	0.13*
Ethylbenzene	16 ± 0		Manganese	0.80*
Methylene chloride	<10		Nickel	4.1*
Tetrachloroethylene	<10		Lead	2.2*
Toluene	30 ± 13		Selenium	0.069
Trichloroethylene	<10		Silver	1.9*
			Thallium	0.065
			Zinc	28*

SOURCE: D. Young.

NOTE: None of the volatile organics have been found in sediments of marine animals. Concentrations are mean ± standard error.

*1977 average from treatment plant monitoring data.

**Sample collected 18 Jan 1972.

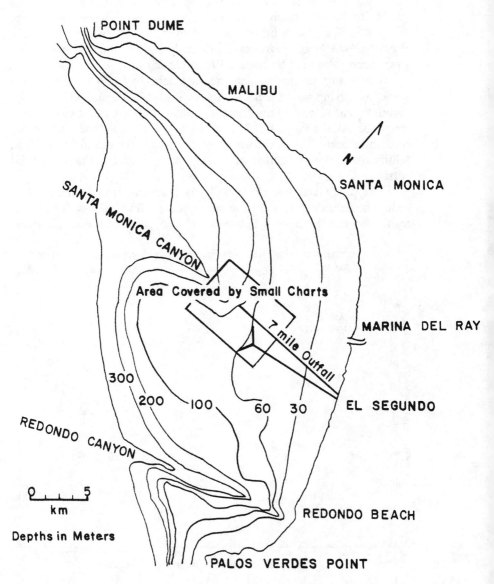

Fig. 13.1 Santa Monica Bay, showing seven-mile outfall discharge point where canyon walls are steep.

We have combined the measurements of volatile solids in the canyon area with our observation (TV, photos, cores) and experience to estimate the amount of sludge remaining near the discharge point. We believe the volume of excess volatile solids (above natural levels) is about 200,000 cubic meters of wet material or between 66,000 and 93,000 tons, dry weight. The total volume is about equivalent to $10 \pm 2\%$ of all the solids discharged over the last twenty-two years. Presumably a larger percentage originally settled, much of which has since been consumed by the sea life in the canyon.

Laboratory settling velocity measurements by Dr. Allan Abati were made by mixing the effluent with seawater at 1:500 and weighing the amounts that settled to the bottom of glass cylinders at increasing time intervals. As shown in fig. 2, one quarter of the solids settled (in the test) in ½ hour. This is equivalent to 2.5 days in the ocean for particles falling forty meters (height of plume bottom above a theoretical flat surface).

The predicted sediment field differs in several ways from the actual sediment field determined from survey data. Nevertheless, it is significant that the differences between the two patterns are qualitatively

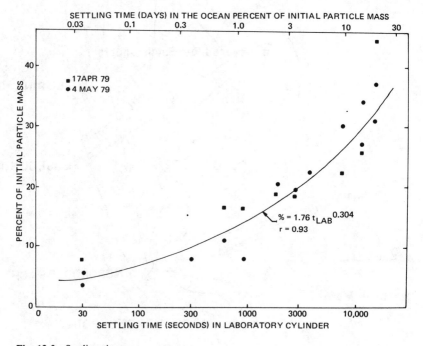

Fig. 13.2 Settling time measurements.

consistent with the observed properties of the currents in the outfall area.

1. The heaviest sedimentation is near the head of the canyon. This is consistent with the up-canyon movement of the near-bottom currents just below the outfall. These currents would carry particulates settling into the canyon toward the head of the canyon.

2. During periods of internal wave activity, the near-bottom currents in the inshore end of the canyon are sufficiently strong to resuspend particulates that have settled in the canyon. These particulates also would be carried toward the canyon head.

3. Low-density waste particulates deposited on the shelf inshore from the canyon are likely to be carried into the canyon during periods when water motion near the bottom is sufficiently strong to initiate resuspension.

All of these processes tend to modify the predicted depositional pattern so that the maximum buildup of effluent particulates will occur near the head of the canyon either by up-canyon movement of particulates in the canyon, or by the offshore movement of particulates on the shelf above.

This theoretical pattern worked out by Dr. Terry Hendricks indicates that 10% of the discharged solids fall within an oval area covering about two square kilometers around the outfall.

The bulk of the solid material discharged with the effluent is very fine particles (96% of the solids will pass through a 0.25 mm screen) of neutral density that neither sink nor float. These remain suspended in midwater for many days, and most are probably carried out to much deeper water offshore by the currents that flow along the edge of the shelf in a generally north westerly direction. Many are probably consumed by sea animals, but in any case they are very widely dispersed.

Particles larger than 1 mm are often of much greater density than the water. These represent about 1% of the solids discharged, and they are assumed to fall out immediately (table 4). There is actually a pile of such material, which looks much like dark grass cuttings, directly in front of the discharge. On close inspection, this sludge material is seen to include watermelon and tomato seeds, plant fibers, and bits of leaves. In addition, there is a considerable number (but very small percentage by weight) of artifacts such as band-aids, cigarette filters, bits of aluminum foil, and small rubber or plastic objects. Particles between 0.25 and 1.0 mm in size make up about 3% of the solids discharged. Most

TABLE 13.4

Size and Amounts of Sludge Particles Settling
to the Bottom Within the First Three Hours* After Discharge

PARTICLE SIZE	EFFLUENT SOLIDS IN SEVEN-MILE DISCHARGE		
	PERCENT OF SOLIDS	TONNAGE PER YEAR	TONNAGE FALLING TO BOTTOM WITHIN THREE HOURS OF DISCHARGE
Greater than 4 mm	.09	60	60
4-2 mm	.24	136	136
2-1 mm	.96	390	390
1-0.5 mm	1.80	1,020	1,020
0.5-0.25 mm	1.12	680	680
Smaller than 0.25 mm	96.00	54,400	2,760
		56,700	5,046

SOURCE: H. Schafer
*About one kilometer from the discharge point.

of these probably fall nearby and doubtless account for much of the sludge that remains in the canyon.

Other factors besides the seven-mile discharge also influence the amount of sludge in the canyon. For example, some small part of the solids released from the far larger five-mile pipe have, at various times in the past (during equipment breakdown or floods), discharged settleable solids onto the shelf, some of which were swept by currents into the canyon.

It has also been suggested that occasional rapid slides of bottom material called turbidity currents, which have been observed in other submarine canyons, transport the detritus and sludge down the steep (10:1) slope into the basin below. We have found no clear evidence of turbidity currents in Santa Monica Canyon, but these could be a factor in the amount of sludge present.

Life on and Near the Pipe

The seven-mile pipe ends in a sled-like steel structure that was used to tow the pipe into place when it was installed in 1957. (Full use for sludge began in 1959). The depth at that point is about 100 meters.

The pipe offers a hard substrate, and sea animals that usually live on rocky bottoms are present there in substantial numbers. The sea anemone *Metridium senile* almost covers the terminal structure. The starfishes, *Astropectin verrilli* and *Luidia foliolata*, are very common on the surrounding soft bottom (especially on the north side of the pipe). The gastropods, *Kelletia kellite, Megasercula carpenteriana,* and *Terebra* sp., were also common on the soft bottom nearby.

The larger rockfish utilize the pipe as a point of reference for schooling; schools of bocaccio (*Sebastes paucispinis*) and vermillion rockfish (*Sebastes miniatus*) are common near the end, and individual bocaccio have been seen cruising along the pipe. Schools of the smaller shortbelly rockfish (*Sebastes jordani*) were also frequently observed above the pipe. Individual cow rockfish (*Sebastes levis*) occurred near the *Metridium,* and Dover sole (*Microstomus pacificus*) were noted on the soft bottom and on the pipe, particularly near the region where it is buried.

The bottom in the vicinity of the seven-mile discharge and at the head of Santa Monica Canyon was examined by towing a television camera along the bottom. Sediments were generally firm on the shelf inshore of the canyon, becoming softer along the pipe and very soft within the canyon where the television sled sank to a depth of at least 0.6 meter in material directly in front of the discharge point that seemed to be largely sludge particulates.

The water was moderately clear away from the pipe and either very clear or very turbid immediately above the sludge field. Sea pens (*Virgularidae*) and sand stars (*Astropectin verrili*), which were common but not abundant on the flat above the canyon, were not observed in the sludge fields. Fishes were more abundant in the sludge field than inshore of the canyon (0.42/sq m *vs* 0.05/sq m); dominant species in the sludge field were white croaker (*Genyonemus lineatus*) and shiner perch (*Cymatogaster aggregata*) with densities of 0.18 and 0.06/sq m, respectively. Dominant species outside the sludge field were longspine comfish (*Zaniolepis latipinnis*), pink sea perch (*Zalembius rosaceus*) and sanddabs (*Citharichthys* sp.) with densities of 0.02, 0.01, and 0.01/sq m, respectively. Pacific electric rays (*Torpedo californica*) occurred in high densities (0.04/sq m) in the sludge field, presumably feeding on schooling fishes. Schools of northern anchovy (*Engraulis mordax*) were observed in the canyon area near the bottom. Both white croaker and shiner perch were observed picking at the bottom, presumably obtaining food. The fact that the water in the canyon above the sludge was clear enough for effective use of bottom television in counting fishes is an indication of water conditions.

Life in the Bottom

The project's 1978 annual report contained charts of the central southern California coastal waters, including Santa Monica Bay and canyon, showing Infaunal Trophic Index (ITI) distribution and biomass of benthic invertebrates (Bascom 1978).

An ITI of sixty (on a scale of 0 to 100) usually represents the lowest "normal" level and thirty to sixty is the region of change (Bascom et al. 1978). Infaunal index numbers below thirty indicate very substantial changes in benthic animal communities. Fig. 3 shows that, in the upper part of the canyon, the number of species is greatly reduced (perhaps as few as 10); Infaunal Trophic Index is often zero.

The bottom animals in the zero to thirty area are largely those of Groups III and IV, which live and feed in the soft mud. These include polychaete and dorvellid worms, two Group III molluscs (*Parvalucina tenuisculpta* and *Macoma carlotensis*), and a Group II mollusc (*Axinopsida serricata*), which usually feeds in the water column. There are about fifteen to twenty species present when the Infaunal Trophic Index is thirty, as it is along the edge of the inner canyon.

The number of individuals present and the biomass in most of these canyon stations is about double the number at control stations. This ecological result is often seen when there is a large specific food supply. The number of species is reduced to the few who can utilize the food; those grow and reproduce exceedingly well.

Life on the Bottom

Trawls and observations made repeatedly in Santa Monica Canyon immediately in front of the sludge discharge show that there is an abundance of fish and invertebrate life above the sludge. Our problem was to find ecologically equivalent control areas with which we could compare the data. In that endeavor, we have only been partly successful because Santa Monica Canyon is unique; it begins well offshore and its bottom has few rocks. Other canyons, such as that at Point Dume, come much closer to shore and are generally too rocky to trawl. For comparison, we have used data from equivalent depths in Redondo Canyon, San Pedro Sea Valley, and off Coal Oil Point and Point Conception.

As can be seen on table 5 there are more species of fish (17.4), more biomass (40.7 kilograms), and larger individual fish (59 grams) in the sludge area than at the stations chosen for comparison. There is an even greater excess over the average numbers found in our 60-meter control survey, where the average was 14 species of fish and 347 individuals per trawl.

The average number of species of invertebrates taken in Santa Monica Canyon is more than the 11 found on the 60-meter control survey; the number of individuals (734) compares with 455 in the control survey.

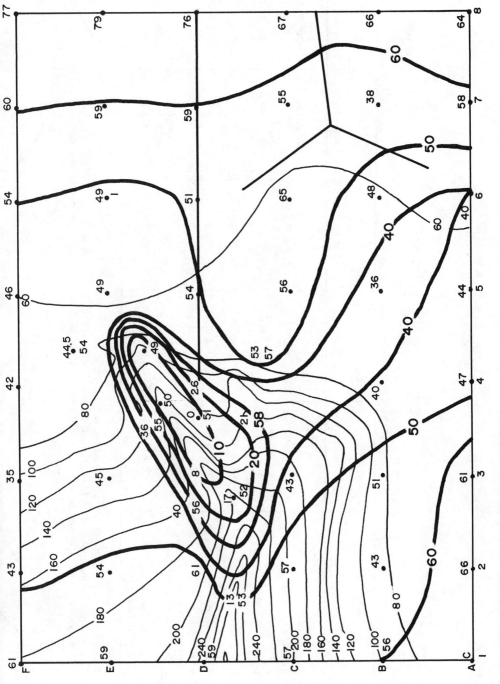

Fig. 13.3 Infaunal trophic index in region of outfall.

TABLE 13.5

Trawl Catch Statistics in Depths of 150 Meters

	Point Conception 1977	Coal Oil Point 1977	Santa Monica Canyon* 1976–1979	San Pedro Sea Valley Canyon 1976–1977	Redondo Canyon 1979
No. of samples	3	3	5	3	2
Depth (m)	152 ± 27	149 ± 21	145 ± 27	139 ± 2.5	164 ± 51
Tow time (minutes)	22 ± 5.7	21 ± 3.8	11.8 ± 4.6	10.0 ± 1.0	10 ± 0
Fish					
No. of individuals	168 ± 102	342 ± 314	691 ± 392	200 ± 189	841 ± 590
No. of species	14.0 ± 3.6	15.0 ± 3.5	17.4 ± 5.0	8.3 ± 1.5	17.0 ± 9.9
Biomass (kg)	6.9 ± 4.2	4.1 ± 3.6	40.7 ± 29.9	6.3 ± 5.6	28.8 ± 12.7
Avg. wt./fish (g)	5.4	12.0	58.9	31.0	38.4
Shannon-Weaver Diversity	1.80 ± 0.20	1.92 ± 0.07	1.69 ± 0.38	0.99 ± 0.60	1.75 ± 0.41
Invertebrates					
No. of individuals	527 ± 664	400 ± 379	734 ± 774	75 ± 70.7	634 ± 276
No. of species	53 ± 27**	25.3 ± 8.6	15.2 ± 6.0	10 ± 2.8	18.0 ± 7.1
Biomass (kg)	7.0 ± 6.6	65.0 ± 23	13.0 ± 8.0	6.5 ± 8.1	6.6 ± 0.63
Avg. wt./invert. (g)	13.3	163.0	4.1	20.8	11.2

SOURCE: A. Mearns

NOTE: The three canyon trawls were only about half the duration of the controls.

*Santa Monica has the most species of fish as well as the most biomass and the largest individual fish.

**Two trawls with many invertebrates attached to drift kelp, sea weeds, and rocks. These are not comparable with soft bottom fauna in other trawls.

Overall, a total of forty-one species of fish and forty species of macroinvertebrates have been captured by trawl at the sludge outfall (tables 6 & 7). Catches are dominated by Dover sole (*Microstomus pacificus*), white croaker (*Genyonemus lineatus*), slender sole (*Lyopsetta exilis*), Pacific sanddab (*Citharichthys sordidus*), the star (*Astropectin verrilli*), the prawn (*Sycionia ingentis*) and sea pens (*Acenthoptillum* sp.). There is no question but that sea life is very abundant above the bottom in the area of the sludge deposit.

Life in Mid Water

Animals caught in the trawls described in the previous section live within one meter of the bottom. Large populations of other species live higher in the water column throughout the Bay.

Santa Monica Bay is very productive and contains an astonishing amount of sea life — at least 70,000 tons — which itself produces about 25,000 tons of fecal material a year (wet weight). The outfalls undoubtedly influence this figure somewhat, presumably enhancing it by providing food and nutrients and perhaps diminishing it by adding toxicants. Alan Mearns of the Project estimates the increase in benthic invertebrate biomass caused by the two Hyperion outfalls (five-mile and seven-mile) at 5,600 metric tons or about 8 percent of the total in the Bay.

The average chlorophyll level in the bay is about 12 mg/liter relative to 4 mg/liter in nearby ocean waters. This indicates the amount of photosynthesis by phytoplankton (diatoms and dinoflagellates), which form the base of the food web. Minute animals, including copepods and larval fish live on these microscopic plants. These, in turn, are consumed by larger animals including anchovies and mackerel, which are in turn consumed by still larger fish, sea lions, and birds.

The outfalls do not appear to have much effect on this primary productivity because the nutrients they release (mainly ammonia nitrogen) usually do not reach the euphotic (sunlit) level of water where photosynthesis can take place and chlorophyll form. If the outfall nutrients reached the surface waters, we would expect to see a local enhancement of phytoplankton that is not observed. Rather the increased chlorophyll seems to come from nitrogen in nitrates from upwelled water that moves into the shallow, lighted water of the Bay when easterly winds push the surface water offshore.

Fish Diseases

Over the past twenty years, numerous diseased fish have been collected from Santa Monica Bay and elsewhere in southern California. Two

TABLE 13.6
Species and Number of Fish Taken in Seven-Mile Outfall Trawls

Species	Number of Individuals					
	1976	1977	1978	1979F	1979W	Total
Depth (m)	128	183	152	149	124	
Tow time (minutes)	20	10	10	9	10	
Species						
Dover Sole	329	232	597	120	177	1455
White Croaker	312	96		1	9	418
Slender Sole		49	177	108	58	392
Pacific Sanddab	28	96	18	11	179	332
Pink Sea Perch	195	4				199
Long Spine Combfish	182	2				184
Speckled Sanddab	93	8				101
Stripetail Rockfish	35	3	3		31	72
Splitnose Rockfish			26	24		50
Shiner Perch	38					38
Plainfin Midshipman	16	1		1	12	30
English Sole	5	13		1	7	26
Rex Sole			20			20
Green Blotched Rockfish	5	4			5	14
Shortbelly Rockfish	2	11				13
Ratfish	13					13
Green Spotted Rockfish			1		11	12
Spotted Cuskeel	3	3	4	1	1	12

Species	1279	529	871	269	509	3457
Blackbelly Eelpout						1
Shortspine Combfish	3	1	10		6	10
Sablefish	8	1	7		1	9
Pacific Electric Ray				1	1	4
Greenstriped Rockfish	4				4	4
Curlfin Sole	3					3
Hornyhead Turbot						3
Blacktip Poacher			3			3
Cow Rockfish	1	1			1	3
Bocaccia		1			1	2
Hundredfathom Codling					2	2
Pigmy Poacher			1	1	2	2
Spiny Dogfish			2			2
Longfin Sanddab			1			2
Gulf Sanddab	1					2
Bigmouth Sole	1					1
California Tonguefish	1	1				1
Squarespot Rockfish		1				1
Calico Rockfish		1				1
Halfbanded Rockfish						1
Shortspine Thornyhead				1		1
Lizardfish	1	1			1	1
Smooth Stargazer	1	1			1	1
Total 41 species	1279	529	871	269	509	3457

Source: A. Mearns

TABLE 13.7

Invertebrate Species Taken in 7-Mile Outfall Trawls

			Number of Individuals			
	1976	1977	1978	1979F	79W	Total
Depth (m)	128	183	152	146	140	
Tow time (minutes)	20	10	10	9	21	
Species						
Astropecten verrilli	718	1188			140	2046
Sicyonia ingentis	340	5	105	13	21	484
Acanthoptilum sp.*		300–500		P	P	400+
Simnia sp.*	50–100	13			2	~90+
Solemya sp.	80					80
Seapen Ul*	50–100					~75
Pandalus platyceros	1		63			64
Crangon alaskensis*	~50	1				~51
Pandalus jordani			44			44
Merridium senile	2	4	33		3	42
Listriolobus pelodes*	1–10	31				~38
Allocentrotus fragilis		1	10		24	35
Armina californica	24	3				27
Parastichopus californicus	12		8		5	25
Pleurobranchaea californica	11			2	7	20
Mediaster aequalis	7	3			1	11
Lironeca vulgaris*	1–10	3			P	~8+
Cancer anthonyi	7					7

Species						
Luidia foliolata		1	4	1	5	7
Calinaticina oldroydii		2				6
Pinnixa sp.*	2–5	1	2	3		5
Parvilucina tenuisculpta					3	5
Lytechinus anamesus	1		2			3
Mursia gaudichaudii		3				3
Cisena californiensis		3				3
Ophiura lutkeni						3
Octopus Prubescens	2			2		2
Nassarius mendicus				1		2
Crangon zacae	1		1			2
Cancer productus	1					1
Tritonia exsulans	1					7
Heptacarpus tenuissimus	1					1
Cerebratulus sp.	1					1
Glycera americana	1					1
Lucinoma annulatus	1					1
Luidia asthenosoma		1				1
Sergestes similis				1		1
Nereidae U1				1		1
Acila sp.					1	1
Metridium sp.					1	1
Total* 40 species	1211	1214	273	24	210	

SOURCE: A. Mearns

*Those species with ranges on approximate numbers were not used in calculations.

types of diseases have received most attention: external tumors on croakers and flatfishes and fin erosion diseases in a variety of fishes.

Present evidence indicates that, in southern California, skin tumors occur commonly in young Dover sole. The tumors are similar to a disease affecting very young flatfishes in Japan, Alaska, British Columbia, Washington, Oregon, and northern California. Prevalence of tumors is low in young Dover sole from southern California compared with those in other species from sites in the Pacific Northwest and Alaska; diseased Dover sole have been captured as far south as Cedros Island in Baja California. The cause of the disease is unknown but whatever it is, it is widespread and has been present in coastal areas of the north Pacific for many years. The disease does not appear to be caused by southern California sewage or sludge discharges.

Few white croaker with lip papillomas have been reported since 1970. This growth has been described as nonmalignant and possibly a response to an irritant or an injury (Russell and Kotin 1957). Because of the relatively wide coastal distribution of the condition and the nature of the growth, the lip papilloma disease does not appear to be related to the discharge of municipal wastewaters (Mearns and Sherwood 1977).

The second type of disease, commonly known as fin erosion syndrome, does seem to be related to sewage discharges although the specific cause is unknown. Many reports by this project (including Sherwood 1978) have described this disease, its epicenter off Palos Verdes, and its prevalence relative to other parts of the world.

Fishes with fin erosion disease have been collected in the vicinity of the seven-mile outfall. The Dover sole (*Microstomus pacificus*) is the species most frequently affected, both at this site and on the Palos Verdes shelf some twenty kilometers to the southeast.

The proximity of Santa Monica Bay to the Palos Verdes shelf and the less frequent surveys of the canyon area have made it difficult to estimate the extent to which migration could be responsible for the presence of Dover sole with fin erosion in the canyon. Fin erosion in Dover sole was found at depths of up to 460 meters off Palos Verdes, although the number of fish affected decreased with increasing depth (52% at 140 meters and 23% at 460 meters). No fish with fin erosion were taken along the isobath between Palos Verdes and Santa Monica Canyon suggesting that this was not a major pathway for the movement of diseased fish from one area to the other.

A trawl sample taken at the end of the seven-mile pipe contained large Dover sole both with and without eroded fins (table 8). A comparison of trace contaminants in the tissue of these and smaller Dover sole suggested that fin erosion was being initiated in bottom fishes in the vicinity of the seven-mile pipe. Prevalence of fin erosion in Dover

TABLE 13.8

PREVALENCE OF FIN EROSION AND SKIN TUMORS IN DOVER SOLE
COLLECTED NEAR THE SEVEN-MILE OUTFALL IN SANTA MONICA CANYON

	1977	1978	1979	1979
Day	29 Sep.	2 Oct.	7 Aug.	7 Aug.
Depth (m)	183	152	149	124
Tow Time (minute)	10	10	9	10
No. of Dover sole	232	597	120	172
No. with fin erosion	38	86	13	56
% with fin erosion	16%	14%	11%	32%
No. with tumors	3	1	1	1
% with tumors	1.3%	0.17%	0.83%	0.58%
No. <120 mm SL	71	28	30	1
No. with tumors <120 mm SL	2	1	1	0
% with tumors <120 mm SL	2.8%	3.6%	3.3%	0

SOURCE: M. Sherwood

sole in the first sample near the five-mile pipe was 4.2%; in the second sample, directly off the seven-mile pipe, 14%; and in the third sample, 4.25 kilometers northwest of the seven-mile pipe, 0%.

Fin erosion in Dover sole and several other species with similar disease distribution patterns appears to be the result of exposure to contaminated sediments. Although a correlation exists between occurrence of fin erosion and elevated levels of total PCB in sediments, a cause-and-effect relationship between the two has not been demonstrated (Sherwood and Mearns 1977). Liver enlargement found in specimens with and without eroded fins collected in the vicinity of the seven-mile pipe is discussed in the trace contaminant section.

Chemistry and Toxicity

Virtually all of the possible pollutants in the city's discharged waste are present in the sediments in upper Santa Monica Canyon, some at high levels. This is because the tiny particles of these pollutants are attached to particles of organic solids that are discharged, about 10% of which fall to the bottom in the discharge area. The result of this is that all pollutants correlate reasonably well with each other and with volatile solids. The correlation with the number of species of marine life is inverse; that is, numbers of species decrease as pollutants increase. However, the biomass and number of animals may increase substantially in spite of the presence of possible toxicants. We do not know which of the nutrients, chemicals or pollutants cause the reductions and/or increases.

An examination of the results of chemical analyses of bottom samples shows that the most convenient and reliable indicator of sewage solids is an increase of volatile solids above the normal levels. Fig. 4 shows the distribution of volatile solids around the outfalls. The natural range of volatile solids, as determined by our control survey for outer shelf depths is 1.8% to 3.8% (Word and Mearns 1978). The flat area above the canyon shows little increase above normal. In the canyon, however, the levels rise abruptly to as much as 33% in a pile of large sludge particles immediately in front of the discharge point.

A similar pattern exists for increased copper in the bottom with levels in the upper canyon reaching as much as 1300 ppm relative to the control range of 2.8 to 31 ppm. Cadmium in the bottom also has the same pattern and range of values. These metals are bound to the particles and there is no evidence that they are biologically available to any great extent. Sometimes the natural metal level in invertebrates will rise to five or ten times normal, but there is no known toxic effect at these levels. The fish that feed on these invertebrates have metal levels that are slightly below normal. The extent to which these pollutants can be made biologically available by microbial action that destroys the underlying organic particles to which they are attached is unknown.

The levels of polychlorinated biphenyls (PCBs) in the canyon sediments are mostly in the range of 1 to 3 ppm (dry weight); levels of PCBs on the shelf immediately above the canyon are about one-tenth that level, and these decrease towards shore (fig. 5). Often fish accumulate PCBs in muscle tissue at about two-tenths the level of that in the sediment. This means that the dry weight in sediments is often about the same as the dry weight in fish muscle. The highest level measured (3.1 ppm) in sediment is only slightly above the permissible level in fish for human consumption.

The concentrations of the total chlorinated benzenes in waste effluents are generally higher than those of PCBs, and they can be detected and measured at the same levels as PCB, or about 0.1 ppb (Young 1978). Around other outfalls, dichlorobenzene has been detected in sediments at about the same level as PCB. Fish taken in trawls of the upper canyon were found to have certain chlorinated benzenes in their livers. Speckled sanddabs were found to have trichlorobenzene and hexachlorobenzene in their livers, each at about 25 ppb. No para- or ortho-dichlorobenzene was detected. Generally the levels of the chlorobenzene in muscle tissue (the edible part) of such fish are about one-twentieth the level in the liver, although we did not make that mea-

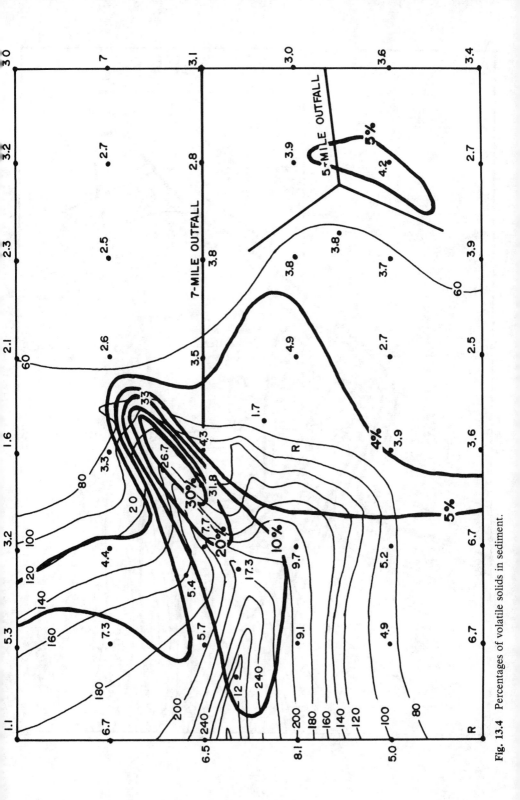

Fig. 13.4 Percentages of volatile solids in sediment.

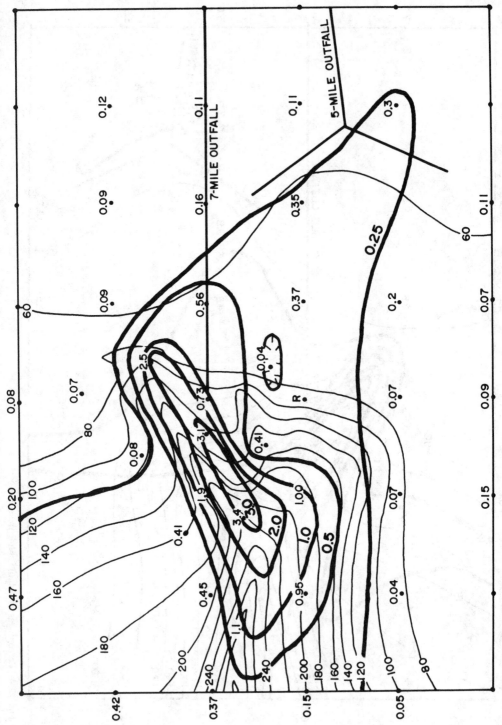

Fig. 13.5 Concentrations of total PCB in upper two centimeters of sediment

surement in the fish from the canyon. When this level is compared with the FDA standard of 300 ppb HCB in meat products, no danger is indicated.

Thus, although the chlorinated benzenes are dominant in effluent and are prominent in sediments, they do not appear to bioaccumulate to any important extent, either here or elsewhere.

Preliminary toxicity studies have been made on the seven-mile effluent using both sea urchin fertilization and embryo development techniques developed by Philip Oshida of this Project, and the effect was greater than anticipated.

Fertilizations were not significantly reduced at dilutions of 700:1 but at a dilution of 600:1, the discharge was slightly toxic to the development of sea urchin embryos. This toxicity continues to decrease as the wastefield moves away and becomes further diluted so that a relatively small volume of seawater is hazardous to tiny organisms. Since the sea urchin reaction is a measurement of overall toxicity, it is not possible to say which of the many constituents in the discharge is responsible.

CONCLUSION

At some time over the past ten years, nearly every scientist on the staff of the project has made some measurement relating to the seven-mile sludge outfall and its effects on the bottom or on sea life. All the important data have been summarized in this paper. The nature of the material discharged, how it is distributed by the currents, where it deposits and why, the extent to which animals are affected, the incidence of pollutants in the bottom, the statistics on fish diseases and toxicity of the plume have all been described.

We find that the area seriously affected is entirely in the canyon at depths greater than 100 meters (300 feet) and that the area seriously involved is about 3 square kilometers — a little more than a square mile or about 0.4% of the Bay. However distasteful it may seem to humans, this sludge-lined canyon attracts sea animals. Although the community structure in this small region is greatly changed, some animals have taken advantage of the situation and proliferated. Others have disappeared, and some rare forms have grown unusually large. The benefits and the detriments of discharging sludge from the seven-mile pipe are mixed.

We hope that any alternative proposal will receive equal study so that any changes made will be an ecological improvement.

Bibliography

Allen, M. J., H. Pecorelli, and J. Word. 1976. Marine organisms around outfall pipes in Santa Monica Bay. *J. Water Poll. Contr. Fed.* 48:1881.

Bascom, W. 1978. Life in the bottom. In Annual Report, 1978, p. 57. Southern California Coastal Water Research Project, Long Beach, Calif.

Bascom, W., A. J. Mearns and J. Q. Word. 1978. Establishing boundaries between normal, changed, and degraded areas. In Annual Report 1978, Southern California Coastal Water Research Project, Long Beach, Calif.

Coastal Water Research Project. 1977. Interim report on sludge. Technical Memorandum 228. Southern California Coastal Water Research Project, El Secundo, CA.

Hendricks, T. J. 1977. Satellite imagery studies. In Annual Report, 1977, p. 75. Southern California Coastal Water Research Project, Long Beach, Calif.

Mearns, A. J., and M. J. Sherwood. 1977. Distribution of neoplasms and other diseases in marine fishes relative to the discharge of wastewater. *Ann. New York Acad. Sci.* 298:210–24.

Mearns, A. J., and L. S. Word. 1975. Hydrographic and microbiological survey of Santa Monica Basin, winter and summer 1974. Technical Memorandum 218. Southern California Coastal Water Research Project, Long Beach, Calif.

Russell, F. E., and P. Kotin. 1957. Squamous papillomas in the white croaker. *J. Nat. Cancer Inst.* 18(6):857–61.

Sherwood, M. J. 1978. The fin erosion syndrome. In Annual Report, 1978, p. 203. Southern California Coastal Water Research Project, Long Beach, Calif.

Sherwood, M. J., and A. J. Mearns. 1977. Environmental significance of fin erosion in southern California demersal fishes. *Ann. New York Acad. Sci.* 298:177–89.

Word, J. Q. 1978. The Infaunal Trophic Index. In Annual Report, 1978, p. 19. Southern California Coastal Water Research Project, Long Beach, Calif.

Word, J. Q., and A. J. Mearns. 1978. The 60-Meter control survey. In Annual Report 1978, p. 19. Southern California Coastal Water Research Project, Long Beach, Calif.

Word, J. Q., and A. J. Mearns. 1979. 60-meter control survey off southern California. Technical Memorandum 229. Southern California Coastal Water Research Project, 1978. Long Beach, Calif.

Word, J. Q., T. J. Kawlings and A. J. Mearns. 1976. A comparative field study of benthic sampling devices used in southern California benthic surveys. Task report to the U.S. Environmental Protection Agency. Grant No. R801152.

Young, D. R. 1978. Priority pollutants in municipal wastewaters. In Annual Report, 1978, p. 103. Southern California Coastal Water Research Project, Long Beach, Calif.

CHAPTER 14

Economics of Sewage Treatment and Disposal Alternatives

Jack M. Betz
Bureau of Sanitation, Los Angeles
United States

Sanitary engineers design sewage treatment plants to produce effluents which can be safely discharged to the environment. The degree of treatment necessary is a function of the discharge location — the receiving water. For example, discharges to small streams or lakes obviously require a higher degree of treatment than a well diffused discharge to the open ocean. To the degree possible, in the face of many non-quantifiable constraints, the function of the engineer is to produce the most cost effective system from the various treatment and disposal alternatives.

SEWAGE TREATMENT PROCESSES

In a simplified way, sewage treatment consists of separating the solids from the liquid, discharging the liquid to a receiving water and doing something with the solids.

The many sewage treatment processes available can generally be grouped into three classifications; primary, secondary and advanced treatment. Solids disposal will be discussed separately.

The objective of primary treatment is to remove pollutants which will settle or float. This is accomplished in settling tanks with detention periods of one to two hours. About 60% of the suspended solids can be removed along with about 30% of the BOD or biochemical oxygen demand. The process is simple, trouble-free and relatively inexpensive.

The principal goal of secondary treatment is to satisfy the soluble BOD which escapes the primary treatment process as well as to remove additional suspended solids. This is accomplished by using the same biological processes which would occur naturally in the receiving water (if it has adequate assimilative capacity) but in an artificially speeded-up manner. The activated sludge process is typically used for secondary treatment. Aeration tanks with a detention period of four to six hours are used with compressed air bubbled through the liquid. These are followed by settling tanks similar to those used in primary treatment. Most of the settled material is returned to the aeration tanks as seed, "activated sludge," to maintain the biological process of converting the colloidal and dissolved pollutants to settleable material. Together with the primary treatment process, secondary treatment processes will remove about 85% of the pollutants. The activated sludge process is much more complex then primary settling and is subject to upsets by all of the factors which can affect biological processes, e.g., overloading, temperature changes, toxic substances and inadequate air supply. Capital cost and operating costs are much higher than for primary treatment.

If even greater removals of pollutants are required, advanced treatment processes can be added to the secondary treated effluent. These methods include physical-chemical processes, for example chemical coagulation and sand filtration such as is typically provided for potable water supplies. Removal of nutrients such as phosphate or nitrate may also be accomplished. The addition of advanced waste treatment will typically increase the overall removal to about 99% of the suspended solids and BOD. As physical-chemical methods, they are moderately complex but are not so easily upset as biological processes. Due to the law of diminishing returns, they are costly in terms of the additional pollutants removed.

The tabulation below summarizes and quantifies some of the factors discussed regarding sewage treatment processes. The cost data uses as an example a 20 million gallon per day treatment plant, which is a size to serve a community of about 200,000 population.

TABLE 14.1

SEWAGE TREATMENT FACTORS

CLASSIFICATION	EXAMPLE	POLLUTION REMOVAL	CAPITAL COST 20 MGD	ANNUAL OPERATING COST 20 MGD
Primary	Settling	50%	$15,000,000	$1,000,000
Secondary	Activated sludge	85%	$45,000,000	$2,000,000
Advanced	Coagulation & filtration	99%	$65,000,000	$3,000,000

MARINE DISPOSAL SYSTEMS

A brief chronology of Los Angeles' sewage disposal history will serve as an example of the development of marine disposal systems.

1873 –   City began construction of public sewers. Population 7,000.

1883 –   Sewage disposed to vegetable farms at the edge of the city. Population 15,000.

1894 –   Sewage piped ten miles to ocean at Santa Monica Bay, discharged through twenty-four-inch outfall 600 feet from shore. Population 70,000.

1904 –   New thirty-inch outfall laid, 940 feet from shore. Population 150,000.

1912 –   Intense complaints of filthy beaches. Several bond issues
to   for improvements defeated. Legal actions. Finally a bond
1925   issue passed for a screening plant and seven-foot outfall, one mile long. Population 1,000,000.

1925 –   New outfall leaked badly because of damage to bell and spigot joints when laid under rush orders in heavy seas.

1925 –   Repairs to leaky outfall unsatisfactory. Most sewage dis-
to   charged about 1,000 feet off shore. Several bond issues
1943   for improvements defeated. Continued beach pollution.

1943 –   Beaches quarantined for ten miles. Population 1,800,-000.

1950 –   Secondary treatment plant constructed. New twelve-foot outfall, one-mile long with eight three-foot diameter ports. Beach quarantine lifted. Population 2,000,000.

1959 –   Converted plant to two-thirds primary treatment and one-third secondary treatment. New twelve-foot outfall five miles long with 166 seven-inch diameter ports. Population 2,500,000.

1980 –   Same treatment and disposal facilities. Population 3,200,000.

This chronology covers a period of about 100 years and, as somewhat of an oversimplification, can be considered as three separate phases:

Stage I, about 50 years. Discharge progressively farther from city, eventually into ocean. No treatment. Increasing complaints but people unwilling to spend money on significant improvements.

Stage II, about 25 years. Under legal action public agrees to spend money on something. Environmental requirements not well developed and technology poorly applied. Program unsatisfactory.

Stage III, about 30 years. Scientific and engineering approaches utilized. Public concerns result in sufficient funding. New technologies developed and applied to solve problems in cost effective and environmentally responsible ways.

As related to use of the ocean for waste disposal, stage I shows no consideration as to the effect on the ocean while stage II reflects at least an attempt to get farther from shore to provide some dilution.

Only during stage III is there recognition of the assimilative capacity of the ocean and the need to match the treatment process with this capacity. For example, the 1950 plant and outfall matched a high degree of treatment with a relatively short outfall. The effluent was discharged at a depth of fifty feet, not very well diffused, about 15:1. The effluent was chlorinated to control the bacterial quality of the shorelines. Primarily on the basis of economics, but also with a view towards improved environmental effects, the plant was modified in 1959 to provide for a lesser degree of treatment but more effective use of the ocean. The mixture of primary and secondary effluent was discharged to a five-mile long outfall at a depth of 200 feet with an initial dilution of about 100:1. This effluent did not need to be chlorinated to meet the bacterial requirements.

The tabulation below compares the capital and operating costs of these two systems, adjusted for comparability to 1980 dollars and a common capacity of 420 million gallons per day.

TABLE 14.2

ALTERNATIVE SYSTEMS COSTS

SYSTEM	PLANT COST	OUTFALL COST	TOTAL CAPITAL	ANNUAL OPERATING COST
Secondary — short outfall	$500,000,000	$15,000,000	$515,000,000	$20,000,000
Primary — long outfall	$150,000,000	$50,000,000	$200,000,000	$ 7,000,000

The substantial economic benefit of using more of the assimilative ability of the ocean is evident with less than one-half the capital cost and one-third the operating cost of the secondary treatment — short outfall system. Environmental impacts of the primary treatment — long outfall system are considered to be at least as satisfactory as the alternate system and generally more satisfactory. (These factors will be discussed in detail by other speakers on the panel).

EFFECT OF REGULATIONS

In addition to the environmental, social, technological and economic factors involved in determining treatment programs, local, state and national regulations have a substantial, and sometimes overriding bearing on the methods and costs of sewage treatment and disposal. Brief mention must be made of the economic impact of two important public laws on the subject of this paper.

U. S. Public Law 92–500 is the basic national legislation covering sewage treatment matters. Among its provisions is a requirement that all public sewage treatment plants must provide at least secondary treatment. This requirement made no distinction between plants discharging to inland waters or to the open ocean. As shown in the previous section of this chapter, the economic consequences of this requirement for coastal communities such as Los Angeles was enormous.

After a long and difficult campaign, the coastal states were able to convince the Congress to provide in Public Law 95–217 for a waiver from secondary treatment for dischargers to the open ocean. Such a waiver would allow the economies described in the previous section to be realized. However, there are very strict requirements which must be met before a waiver will be allowed. All of the major dischargers in California have applied for a waiver as well as many other ocean dischargers throughout the nation. The U. S. Environmental Protection Agency is currently evaluating these waiver requests.

SLUDGE DISPOSAL

The residual solids which are removed from sewage are termed sludge and must also be disposed of. The factors involved in the treatment and disposal of this material are becoming even more complicated than those involved with the liquid effluent. This is particularly the case for urban areas where land disposal options are limited. Time and space will not allow a detailed discussion of sludge treatment and disposal in this paper but a brief overview may put the matter in perspective with respect to marine disposal.

It is, of course, obvious that sludge can only be disposed of into the land, the water or the air. A substantial amount of sludge is currently being disposed of in each of these; by crop utilization or land filling, into the ocean or into the air by incineration. Disposal into the ocean is accomplished generally by barging and discharging slightly below the surface. Los Angeles is an exception in discharging by a seven-mile pipeline at a depth of 300 feet at the head of an extensive submarine canyon.

The Federal Government clearly intends to prohibit any discharge of sludge to the ocean and all current dischargers are on timetables to cease. Los Angeles recently signed a consent decree to cease discharging to the ocean by July 1, 1985.

Because land disposal could not be accomplished in an urban area such as Los Angeles, the alternative selected will include a mechanical dewatering process such as centrifuging, a multiple effect evaporation process, followed by some type of pyrolysis. The technology is complex and expensive.

A comparison of the costs of the existing marine disposal system and the alternate disposal systems is presented in Table 3, estimated in 1980 dollars for equivalent capacities of 260 tons/day.

TABLE 14.3

COSTS OF ALTERNATE SLUDGE DISPOSAL SYSTEMS

SYSTEM	CAPITAL COST	ANNUAL OPERATING COST
Ocean disposal	$ 7,000,000	$ 1,000,000
Thermal process	$200,000,000	$15,000,000

A discussion of the relative environmental impacts of these two systems is also beyond the scope of this paper. Suffice it to say that they have been long debated with strongly held opposing views.

CHAPTER 15

The Southern California Bight –
Municipal Sewage Discharges:
A Study in Ocean Pollution
Management

THOMAS C. JORLING
Williams College
United States

The history of the Southern California Bight is an instructive, albeit complicated, environmental case study. It represents a window into the general history of efforts to control the discharges of pollutants from publicly owned treatment works; the difficulties associated with de- tecting and assessing ocean pollution — even in near shore areas; and the difficulties associated with reversing policies concerning waste of water and nutrients even in an acutely water short area.

It is necessary to provide a brief historical framework in order to understand the current governmental/regulatory status of pollution con- trol efforts applicable to the Southern California Bight. Until 1972, defining somewhat arbitrarily a chronology threshold, the prevailing perception of municipal waste by all parties was characterized by con- cern only with oxygen demand, suspended solids and fecal coliform. And prior to that time those pollution parameters were regulated, if at all, to conform to the "assimilative capacity" of receiving waters. Given ocean waters and monitoring and analysis limited to those generalized parameters of pollution, it is understandable why ocean outfalls were attractive alternatives to waste treatment. The 1972 Federal Water Pol- lution Control Act (FWPCA) or Clean Water Act and some parallel efforts of the State of California changed all that.

By 1972 the notion that pollution control which depended on an ability to predict the effect of discharges on receiving waters was de- termined not to achieve pollution control but to in fact reward, if not encourage, pollution. The earlier program included a calculation of "the assimilative capacity", which can be defined as that volume of pollutants which could be processed, treated or otherwise gotten rid of in the receiving waters while still maintaining the acceptable use of the water. For ocean water even the notion of "use" didn't make a great deal of sense. The calculation of such an assimilative capacity assumed knowledge of the structure and function of the aquatic ecosystems over long periods of time, which simply did not exist and will not exist into the indefinite future. Consequently, assimilative capacity became a rather rough negotiated estimate often made by lawyers and engineers, (and cost conscious accountants) certainly not by biologists, of what waste treatment services could be rendered by a particular reach of water.

In addition to concepts such as "beneficial use" and "assimilative capacity," the control program required further logical incongruities such as the provision of mixing zones which are defined as those areas of greater or lesser distance around an outfall source in which measurements are not taken and environmental standards do not apply. Mixing zones were used strictly for the purpose of allowing another layer of negotiation and compromise, always with the burden of proof on the government, the public, and the environment. Ultimately, mixing zones were defined to accommodate whatever level of performance that was going to be installed before discharge.

The whole program assumed that matter and energy moved in linear pathways. It was fundamentally opposite to the ecologist notion of keeping matter and energy within as tight as possible circles or cycles.

The 1972 FWPCA applied a base minimum level of performance to all publicly owned treatment systems. The so called "secondary treatment" requirement to be applied regardless of receiving water. More stringent limits could be applied if necessary for receiving water quality. Unfortunately, the EPA definition in 1974, contrary to the intention and expectation of the act, was limited to the same trio of "conventional" pollutants as were needed to control municipal pollution pre-1972; BOD (Biological Oxygen Demand), TSS (Total Suspended Solids) and fecal coliform. Thus, the promise of the 1972 Act to recognize the specific biological, chemical, and physical parameters of waste streams was partially frustrated by the regulations. In fact, even today, most POTWs haven't the foggiest notion of the specific chemical composition of their liquid waste streams of sludges resulting from treatment.

Even pre-1972 – naturally enough, since the primary focus was oxygen demand and suspended solids – many coastal communities made the claim their effluents were not causing any effect on coastal waters. The Congress, however, made no exception for them. Coastal discharges, like all other POTWs were required to meet the technological norm of secondary treatment.

In concert with the state of California – which was developing its "Ocean Plan" to regulate ocean discharges – these coastal communities deferred the necessary construction, in part, because they were anticipating and actively seeking another bite of the legislative apple which they got in the 1977 FWPCA amendments.

Section 301(h) added by the 1977 Amendment to the Clean Water Act established a mechanism by which ocean discharging coastal communities may seek from EPA a modification of the generally applicable secondary treatment requirement. This is not the place to discuss specific disputes concerning the legislation or promulgation of the implementing regulations. Those are interesting issues but not directly relevant to this book.

What I would like to touch on are some features and limitations of the so called 301(h) program which may have broader applicability in the area of ocean pollution.

The advantage of the 1972 law and its conceptual underpinning is its transfer of the burden of proof, of the need to show harm before remedial action is taken, from the environment, public health and the government. This concept is firmly rooted in two lines of thought, both based on ecological principles.

The first is the fact that ecosystems, are complicated – certainly more complicated than we can presently know. Thus confidence in predictive capability for ecosystems is faulty and they should not be intentionally degraded to the point of manifest harm because recovery, even if theoretically possible, may be extremely expensive.

The second line of thought is the reality that our industrial-social-economic system, if it is to remain stable in its structure and function over time, should be managed so that as far as mans' activities are concerned there is no alteration of the biological, chemical or physical integrity of the biosphere. This flows from the recognition that these ecosystems have evolved through constant interaction of those characteristics over millions of years.

The objective of the concept is a maximal patterning of human communities upon natural biogeochemical cycles and a minimum departure from the geological, or background rates of change in the biosphere. Framed another way, the objective is to move from linear pathways in the movement of matter and energy to circular pathways.

In the words of Section 301 of the Federal Clean Water Statute itself:

> The Administrator shall encourage waste treatment management which results in the construction of revenue producing facilities providing for,

(1) the recycling of potential sewage pollutants through the production of agriculture, silviculture, or aquaculture products, or any combination thereof;

(2) the confined and contained disposal of pollutants not recycled;

(3) the reclamation of wastewater; and

(4) the ultimate disposal of sludge in a manner that will not result in environmental hazards.

This is particularly relevant to the case of Southern California and 301(h) where fresh water is so scarce.

The shift of the burden of proof that occurred in the 1972 Act that I mentioned is, on its face, most directly called into question by the 301(h) program. The question can be asked — did Section 301(h) restore the pollution control strategy to the pre-1972 status, as it related to municipal discharges? My summary answer is no, but it has some ominous implications in that direction especially if implementation goes awry.

Before a modification of the secondary treatment requirement can be allowed, the structure of 301(h) requires the applicant to show that a less than secondary effluent will provide for the "protection and propagation of a balanced, indigenous population of shellfish, fish and wildlife." The problem is that ignorance is often viewed as "no effect" so lack of information, which is so prevalent, is improperly and wrongly interpreted. Since we know so little about the structure and function of the biota of near shore ocean waters, this could lead to decisions or findings of approval, especially under the acute pressure of saving money in the short term.

This concern is especially troubling when we realize how little we know of the specific chemical composition of municipal waste streams. As I have pointed out, because of the regulatory sanitary engineering history, municipalities have been looked on as only discharging oxygen demanding — but otherwise benign — organic material. But, here somewhat surprisingly, is where 301(h) is having the effect of driving the municipal control program faster and farther in the direction established under the 1972 Act by requiring that specific attention be given to the identity and chemical character of pollutants. The 301(h) emphasis on pretreatment, on toxics and on monitoring will force this orientation. To its credit the California Ocean Plan also moves the program in this direction.

This direction should be reinforced by the EPA–Municipal POTW Toxic Program now underway. This program not only involves sampling representative POTW waste streams for a broad array of chemicals (of the 129 priority pollutants, all but 29 have been found in the average

POTW waste stream) but also examines the efficiency of various treatment processes in removing these chemicals from the waste streams. The 301(h) program will remain seriously deficient because of our inability to know how specific chemicals enter to biogeochemical cycles, especially synthetic organic materials, and what is their effect, especially over twenty and thirty year periods. Errors will be made — the question is in which direction will those errors be made.

In the case of Southern California, in fact because of it, 301(h) is structured to emphasize the reclamation-recycling orientation of the 1972 Act. The conditions on no new discharges and cross references to the reuse provisions of the act make it clear that discharging fresh water into the oceans, especially from water short areas, does not make good sense.

The prudence of this orientation seems confirmed on reading in the April 1980 issue of the Journal of the Federal Water Pollution Control an article by David Argo of the Orange County Water District on the "Cost of Water Reclamation by Advanced Wastewater Treatment." The conclusion:

> These data, when compared equally, demonstrate that the least expensive water available is reclaimed wastewater. The Orange County Water District believes that the future growth and economy of Southern California will depend to a great extent on the continued development of wastewater reclamation.

The policy of recycle-reuse is reinforced in the Southern California area because of water shortage, but the limitation of our present and I think future knowledge, on the character of pollutants in municipal waste streams and their movement into the structure and functions of ocean ecosystems should give us all pause whenever coastal outfalls are located.

The apparent size and volume of the oceans must not be allowed to allow any society, especially ours, to ignore the fact that the oceans are part of the biospheres. Nor can we ignore, usually in the name of "we can't detect significant effects," the immense complexity of coastal ecosystems and their essential linkages to human well being. There are no "sinks"; fresh water should be recycled, nutrients reused and toxics controlled at the source. 301(h) should not be interpreted as an abandonment of these principles but rather as the provision of a transition program to achieve them.

CHAPTER 16

Commentary
Goals for Waste Treatment

TUDOR DAVIES
Environmental Protection Agency
United States

My experience has been such that I do not want to see any degradation at all in estuaries and coastal zones. I have worked with Chesapeake Bay and the Great Lakes, and I have seen the two viewpoints that have been presented at this conference in action many times. One viewpoint says there are plenty of fish and that toxic materials are not up to a specific level in these animals. There may be eutrophication going on, but that is not a bad thing because it is not costing us very much and we do not have major harm.

The second viewpoint, on the other hand, is that there is not a long history of understanding the system. Science says that we are not able, really to predict the consequences of these discharges into the system, and, therefore, we should be conservative and back away from that.

Another issue is the economic cost, which is difficult to back away from. In dealing in the Chesapeake Bay and in the Gulf Coast areas, I have seen local residents and particularly local politicians try to back away as far as possible from the increased economic burdens of secondary and tertiary treatment even though this has displayed rather beneficial improvements.

I recently spent time with the Thames Water Authority, which is the success story of restoring an estuary. Sections of the river had zero dissolved oxygen, and certain fish had not been found in the river even in historical time. The Thames Water Authority decided to make the river non-noxious again so that it would not smell as bad as it had in the past. So they went to a waste treatment system that projected certain municipal growths and certain industrial growths. They were lucky because what they experienced was a decrease in population and a decrease in industrial growth. They put a lot of oxygen back into the river, and fish are starting to come back. I do not know what the ecosystem was like before, perhaps no one does, but at least something has come back; there is life.

What they have done, I think, reinforces what Thomas Jorling has said. They have a single agency which worries about water resources and treats water resources as a unit. The Thames Water Authority provides drinking water for the population of the basin and covers the whole aspect of municipal treatment. The authority operates and regulates the treatment plants, and they are paid for in that way. I feel they are balancing the two issues that we have talked about here of water reuse and treatment. Perhaps those are the issues that we should debate.

We have been talking about a system in which municipal discharges have minimum possible impacts. I agree with Dr. Jorling that our experience in testing the toxicity of municipal discharges shows that they are horribly toxic. They are many times worse than many industrial discharges. I am not sure really what we are opening ourselves up to here unless we have major pre-treatment before we put things into municipal treatment plants.

I would like to see a discussion of how predictable change is in coastal waters, and also energy costs and water re-use costs.

Discussion

Bascom: First of all, let me address several of the items that have just been mentioned, that we do not know about what is discharged or about biology. I submit that every year for the last eight years or so our project has published, in considerable detail, exactly what the Southern California outfalls discharge, in addition to what comes out of the harbors, rivers, aerial fallout, and such; this includes EPA's list of 137 pollutants. We know exactly what is discharged. As for biology, I spoke about that in my discussion (see Chapter 13).

The questions that should be asked are: What is this relative to? What are the normal conditions out there? What has changed? I reported that as fairly as I could. I did not say whether it was good or bad, I simply said this is what the conditions are at the moment. Basically, what I showed was that whatever the problem is at the seven-mile outfall, it is concentrated in 3 square kilometers at the bottom of a canyon, which is at least 200 meters deep and 7 miles offshore. What you should ask is, if you do not put it there, where do you put it instead.

Now, let me briefly give an approximation of what Los Angeles would do instead of discharging, at an additional spending of $200 million down and $15 million yearly. The city would go to a drying process which, in effect, puts a good deal of material into the air, where Los Angeles already has a problem. The residual material would then be trucked some thirty miles over land (perhaps twenty depending on the class of landfill that it goes to) and spread out on the ground in an area of roughly three square kilometers above the San Fernando Valley where there is a reasonable chance that the sludge can become involved in the water table. The EPA ought to be an agency which is equally concerned with land, water, and air, and it should also be concerned about the rather substantial energy costs involved. Why they get into water reclamation at this time I do not know. You must ask a few more questions about water reclamation and think about the process by which water is reclaimed. When you start cleaning up some of the water, you raise the concentration in the other half which you cannot clean up. Now, the ultimate limit on all cleanup is the cost of taking the salt out of the water. Soon you have a situation in which you might as well be de-salting the sea. If your cleanup sufficiently raises the level of pollutants, it becomes impossible to deal with it and, within present regulations, a much more difficult problem than we have now.

As for the question of prediction of effects caused by discharges, our laboratory has already made predictions — some of them were made four years ago. For example, in Santa Monica Bay our two most senior biologists spent a good deal of time and they estimated that about eight thousand tons of excess sea life exists there now. That is, there are eight thousand tons more than would be there under normal conditions because of the feeding from the outfalls. One prediction you can certainly make is that if the outfalls cease to exist you are going to kill eight thousand tons of animals. Whether that is good or bad is up to you. You can also say something about the burial rates of material in the bottom and the rates of erosion. The one thing that you really can not predict is the results of episodic events; that is very difficult.

The one hundred-year storm or a tsunami or some other grand event makes predictions difficult; other than that, we can probably make predictions fairly well.

Betz: I described what I thought were the three stages that Los Angeles has gone through in disposing of its waste products. The first stage was to get it out of sight, out beyond the edge of town. The second stage was to do a little something better with it although we did not really know what to do. The third stage is what we are now doing; that is, applying increasingly better scientific knowledge to solve the problems — to deal with it in some sort of a cost-effective manner. I hope the fourth stage is not doing what some legal edict says to do on a uniform basis throughout the United States. I would hate to see us take that approach, or something similar.

It is tough spending the amounts of money that are required for a situation like ours. It is eased a little bit if the population and the engineers and the politicians, i.e., the local politicians, feel that it is necessary and desirable. It becomes almost impossible if the local population does not support these edicts from on high. Certainly, we would be a lot more comfortable if we were continued to be guided by scientific, well-founded information rather than by some of the Congressional actions that have guided us in the past.

Jorling: It's always difficult to use the South California Bight area as an example. I mentioned in my prepared remarks that it does provide a good one, but at the same time it is atypical for two reasons. First, it has been the subject of more research (i.e., the bight, the outfalls, and the general collection of five systems) than all the others around this country combined, multiplied by several hundred, so that it has an atypical characteristic to it. The second atypical characteristic is that going into a California waste treatment system is like going into a nuclear submarine; while other waste treatment facilities are more akin to the ironclad Monitor. There is a major difference in the character of the waste treatment plants in the East and the West. They have moved much further and faster in the direction of professionalism in the West than in either the East or in the midland. Willard Bascom should feel proud of his work and so should all the communities that have been involved.

Bascom: Local money amounting to $5.5 million has been spent on this project.

Jorling: The situation in Southern California has been well studied. However, not all of the biologists that read the data agree with Mr. Bascom's assessment that there are relatively benign ecological effects. But that dispute does not bother me.

What I want to emphasize in my remarks is not that we cannot place certain materials in the ocean, but rather that we, as a society, must identify all of the materials that we move about. We need to know what those materials are and attack them where they can be attacked. I don't know what the present status of the pre-treatment program is in Los Angeles, but I suspect it is very poor and should be a lot better. I would be much more inclined to support movement of materials from Los Angeles into the ocean, if I knew that they have a good, aggressive pre-treatment program. On a similar basis, I would support movement from other communities, of their material into fresh receiving waters or into ocean receiving water. If you have any doubts about the status of knowledge of the materials that move into municipal waste systems, go to your local system — big or small — and ask them when they last did a full array chemical analysis of that system. Do they know what the materials are or where they are coming from? Los Angeles does not have the problem very frequently, but if you consider an eastern system like *Providence,* which flushes raw, untreated materials thirty or forty days a year, I suspect, into this estuary, you have some idea that we are not as far along on a national basis as is Southern California.

Many of my comments in 301(h) generally, are directed to the national view of it, not just to the application of it in Southern California.

With respect to fresh water, the last time that I looked at the data, the water leaving the diffuser ports in the Los Angeles outfall had a lesser salt content than water that comes into the Los Angeles drinking water system. It is diluted with Sierra Mountain water from Northern California so that it does have a tolerable level when it reaches the citizens of Los Angeles. Effluent is a freshwater resource, and it will be a long time before we desalinate ocean water off the Southern California Bight. However, I think the issue should be looked at. I know in San Diego and in Orange County reclamation is the way of the future; whether it is in Los Angeles will be determined when the water contracts are renegotiated with the reclamation agencies in the mid-1980s.

Comment: Mr. Bascom, first, your laboratory is one of the few in the world that has shown an association of fin rot disease with sewage outfalls. I was wondering if you have progressed in your studies to find a cause-effect relationship with respect to what material was actually contributing to this condition?

The second question concerns the kelp beds. In the days when Wheeler North did some very intensive work in that area, he developed a theory that the sewage was causing population explosions of sea urchins because the amino acids and other organics present in the sewage were being utilized by these urchins which were grazing on the young kelp and not allowing them to propagate. Have you done any further work to either confirm or to show other cause-effect relationships there?

Bascom: Let me deal with the sea urchin business first. Actually, there are approximately a dozen possible reasons why the kelp beds could have gone away including, turbidity in the water (which may or may not be related to the outfall), the sea urchins, DDT, great storms, a general depression in kelp along the coast, or a warm water year.

When the kelp beds started to come back, we looked intensively at all that again, and we never did decide either why the kelp beds went away or why they came back. It is probably a result of one of those reasons — the sea urchins are only partly responsible, if at all. A factor may be the large Japanese market for sea urchin gonads these days. We have loads of people harvesting sea urchins. However, we really can not tell whether that is helpful or not in the whole business.

As for the fin erosion disease, of the fish that are in the canyon at Santa Monica, the amount of fin erosion in two species of fish ranges between 11% and 25%, depending on the year — since we have measurements extending over about five years. That figure is for Dover sole, which is out of its territory in Southern California anyway and does not grow very big. They are basically a Northern California-Oregon fish, where they do grow to commercial size. As far as I know, none are ever caught or eaten in Southern California, because they never grow more than about fifteen to twenty centimeters long.

Dover sole does have fin erosion syndrome, as we like to call it. We have investigated about seven possible causes of that disease and are down to about two now. One of these is the possibility these fish are picking up PCB. It then gets into the liver system and eventually these fish go through a sort of molting and non-eating stage at which time the lipids in the body get depleted and the PCB is released into the nervous system. It appears that the extremities of the nervous system (the capillaries) may be restricted and the loss of blood supply to the extremities causes the fin erosion. This is just a hypothesis.

The other possible cause is DDT. We see this disease mostly off

Palos Verdes, which is noted for the DDT in the sediments, and not in Santa Monica Bay. The DDT, which has dropped to a tenth now of what it was about seven years ago, killed off the microcrustaceans, which are the favorite food of the fish there. Because certain kinds of fish particularly need the chitinous material found in those microcrustaceans in their diet, we are pursuing another hypothesis at the moment: that fin erosion is the effect of a nutritional deficiency caused by the temporary absence of microcrustaceans.

The microcrustaceans are now coming back, and we see the Dover sole shifting their diet. They have changed over to worms, and now they are shifting back. We will have to wait at least a year or two to see if they will be rid of the disease.

Schneider: The complexities and the interlinkages that we have in the marine ecosystems are complex; the fluctuations in the kelp beds illustrates that the linkages are being affected. Part of the complexities that we face in managing these systems is not understanding the mechanisms of these linkages.

Bascom: Are these things that we can do anything about? The biology is a terribly complicated business. Somebody asked me how long is it going to take to understand the ecology of Southern California, and I estimated one thousand years. That is because you really do not have much of a grasp of these long-term cycles and natural processes. I don't know whether it is necessary to know those things or not to solve the essential problem that is before us, which is: what is the best thing to do?

Schneider: To elaborate on the question just a bit more, on a broad regional basis: what is the incidence of fin rot in the area now, say in the Los Angeles Bight?

Bascom: Fin rot has existed in the Southern California Bight as far back as we have samples, but at a much lower level. We have samples taken at San Miguel Island in 1910, and there was fin erosion syndrome then. Fin rot is not like aquarium fin rot, which is a bacteriological problem, but there is no question that the disease has been around for a long time.

We also looked extensively at tumors. Throughout the bight, the existence of tumors in quite a number of species of fish, mostly rock fish, but to some extent flatfish, is about 0.5% across the board; it has nothing to do with outfalls — there is no connection that we can determine.

The fin erosion syndrome, however, is connected with outfalls, for the reasons that I have indicated, and generally speaking, about 90% of it occurs along the Palos Verdes shelf where DDT was the highest. This led us to these hypotheses about DDT and PCBs.

In most of the rest of the bight, fin rot is negligible.You can always find an isolated fish, and you are never quite certain whether he got it someplace else, or whether he swam a long distance from an outfall. On the other hand, I don't feel it is terribly important to know the answer to that in all its details. That doesn't seem to be very relevant to the question.

Schneider: Fin rot is a good indication of chronic and sometimes acute events. Lucien Laubier pointed out (see Chapter 5) that the fin rot that evolved after the *Amoco Cadiz* oil spill, and there was a very marked increase, could almost certainly be tied to that pollution event.

The same type of thing is, needless to say, known by others in the New York Bight area, which is again associated with anthropogenic fluxes into the area.

Bascom: I should mention that it is a worldwide phenomena. We have just completed a study in which we sent out approximately three hundred queries around the world, to places we thought might have it. Although we did not get a very high return, we received about thirty or forty responses; at least half had recognized it locally and several said they would investigate. Our lab has also looked at it in Seattle and New York, as you probably know, so we have some understanding of those areas also.

Comment: Mr. Bascom, given the eloquence of your argument about how much is known about the sewage spills, the effects of the offshore dumping or offshore disposals, as well as the extraordinary expenses involved in trying to use alternatives as Jack Betz mentioned, why did Los Angeles sign the court decree to give up doing this in 1985?

Betz: Los Angeles is faced with about $500 million of improvements required for its sewer system beyond that I have indicated. For example, we have a new sewage treatment plant in the San Fernando Valley that was designed in 1973 and is awaiting a grant offer. The estimated cost in 1973 for that treatment plant was about $30 million. The cost in 1980 is estimated to be about $119 million for that same plant. The term "hostage" has been used in connection with that particular grant offer. For example, all of the time that the plant was ready for a 75% federal and 12.5% state grants, which is 87.5% of the payment for that plant, no grant offer was forthcoming because of Los Angeles's continued desire to maintain its sludge disposal via the sludge outfall.

That is an example of the type of money that is involved. The other $400 million is for improvements in the collection system, the pumping plant, etc. No federal grant offers were being made in that regard because of the refusal to adhere to a sludge-out schedule.

Our city council looked very closely at the possibility of going it alone and making our own payments. It was kind of a toss-up economically — maybe it would have been better. Beyond that, there is no assurance that we would not have had injunctive action taken to prevent the city from going its own way and discharging as it felt was proper. So it was with an agonizing re-appraisal that the city signed the consent decree.

Knauss: It seems to me one of two things must have happened in the EPA to push this issue: either EPA looked at what to do with the sludge as a total ecosystem approach and made a decision that Los Angeles was wrong or EPA is so organized that a narrow perspective was viewed by EPA, ignoring the total ecosystem problem of how you deal with something you have to get rid of. Can anyone tell me which was the approach that was used by the EPA in this matter?

Jorling: Both of these folks are in R&D and since I was formerly involved in this negotiation, I will try to answer. First of all, there is not a shared view of the science of matters, as suggested by Mr. Bascom. This is one area in which the State of California has been a co-prosecutor with EPA in achieving the termination of the disposal of sludge for about five years. There has been a very persuasive argument from a local political viewpoint (it is a money problem). Another problem is that the view of these materials in the ocean is not shared — either by the State of California or by EPA. It has been a long history.

I also think that the cost estimates here for the thermal process alternative are high. There may be other alternatives. For example, in the New York metropolitan area, in two cases, electricity for plant operation will be supplied by generating power from the combustion of sludge. Whether or not that is possible here, I do not know, but energy cost is an increasing problem with the operation of maintenance at these plants.

I no longer have at my disposal the detailed facts on the history of this negotiation. It has been a long and elaborate one. There should be a more environmentally sound way of managing Los Angeles' sludge as there is elsewhere around the country.

If you look at this problem in the context of the oceans, in our present economic climate, oceans as disposal sites are beginning to look very attractive. For instance, the fourteen publicly owned treatment works (POTWs) in the New York metropolitan area, which will be generating 1,457,000 tons a day of sludge with the kind of cost patterns we have seen here, are looking to the oceans in a very serious way now, even though everybody thought we had that problem under control.

Similarly, when you look at chemical production generally in this country and you look at the increasing stringency with which environmental controls are being placed on water, air, and land, you see that the oceans are going to be looked on very favorably by many people as the solution to their problem. It is going to take a great deal of will to keep that from being a cheap alternative. Because I have my fears about that, I want to insist on knowing what we are talking about and what we are doing before we commence any action.

We need to take a holistic approach to environmental management, but it will be expensive. We are seeing in the case of Los Angeles, the New York metropolitan area, and any metropolitan area, what it takes to provide for the needs of the populace over long periods of time. We have not done very well with that problem; we are just now starting to deal with it.

Comment: Willard Bascom alludes to scientists who disagree with the assessment of the Los Angeles-Southern California Bight with respect to how badly it has been damaged by these outfalls. Who are these scientists? What information are they working with? Mr. Bascom does not have a monopoly on information about the ecosystem in the Southern California Bight, but he has a big piece of the action. Now, who are these scientists that are opposing this, and what is the basis of their arguments?

Jorling: There are two dimensions to this problem; one is the liquid effluent problem and the other is the outfall disposal of sludge from the limited treatment facility, which Mr. Bascom discussed.

With respect to the liquid effluent, which is the subject of the 301(h) application, scientists are involved in the review of data supplied by Mr. Bascom and others from the EPA, NOAA, U.S. Fish and Wildlife Service, as well as those who will participate in the public proceedings that will accompany each of those decisions.

With respect to the sludge disposal, I can not recall the people who have evaluated the sludge and its long-term anticipated consequences and formed an opinion that it is not a good thing to continue. Such opinions existed at both the state and federal levels, but I can't recall the individuals.

Kester: It seems as though much of Dr. Jorling's argument was based on not being satisfied with the presently available information. Other than just having chemical analyses of what is being put out, what else would one want to know to make these decisions? One of the factors that comes out very clearly, is the cost of our ignorance — if we can define what that ignorance is.

Jorling: One of the things that science continually gets caught up in doing is justifying an existing practice, rather than coming up with a better practice. In other words, going back to source control would be a much more effective way, because you do not have to worry about continually determining whether or not there is an adverse effect out there in the environment. That is a game you never win, because there is always more environment for which to determine adverse impacts. Concerning the kind of research that would be necessary, everyone could have his own set of studies that would produce the ability to predict. My feeling is that there's not enough money that would go into science that would produce that. Therefore, we should come up as a matter of public policy with a better way of managing our waste, one that does not say, dump it someplace unless there is an adverse effect.

What are the effects? Concerning Mr. Bascom's study in the South California Bight, I do not know how much research they have done on fish behavior, fish communications, or on those areas where evidence is increasing that fish communicate by the chemical balances in the water. All that would be nice to have before you make a judgment that would effect an evolutionarily stable environment. Whether we can make those kinds of empirical judgments, I leave to this kind of audience.

Bascom: I think the question really is should more scientific knowledge be available? My answer to that is that there always ought to be more scientific knowledge available. I have never heard of anybody that thought otherwise. Concerning source control, everybody I know thinks that it is a good idea. You do the best you can with it, but sometimes you reach a point where you can not do much more in that direction. Regardless of the dump in which they want the city to put the sludge, there will be an impact on the environment and its biota. One way or another the water table will be affected, and nobody is studying that.

I do not favor ocean dumping particularly; what I favor is common sense. I feel you ought to figure out what is the best thing you can do and do that, whatever it is. If it turns out when you have studied the land that there are really less difficulties on land than in the sea, you ought to put the waste materials on land. However, I do not think anybody knows anything about that now — certainly nothing comparable to what we know about the sea.

Comment: I am astonished by this discussion. I have never been in a room with so many biologists who were so casual about pollution. I think it is unusual, but I do not object to it. I do think we ought to remember that we have been hearing calls for public information and education, for planning and for doing the best thing. No one can disagree with wanting to do the best thing or wanting to inform the public, but it seems clear from this discussion that it matters who informs the public, because no one agrees. No one agrees on the science, no one agrees on the cost, and surely no one will ever agree on the benefits if they let a Congress loose on the benefits. What does it mean to inform the public? It really does not mean very much at all.

Comment: I would like to comment on Thomas Jorling's remarks about point source controls. I think we all agree that is what we would like to do with sewage disposal. My impression, based upon people to whom I have talked, is that if you do all the point source control you can you may only reduce many hazardous metals by about 40%. The amount of cadmium that comes to sewage sludge is still above what was, until a few weeks ago, considered a toxic level. Preliminary concentration levels suggested for cadmium were such that no amount of sewage sludge that was going to come out of the best sewage treatment plant in the United States, would meet the requirements as for toxic waste. But now, I think all but one or two of the major sewage plants in the United States can meet the most recent requirements for cadmium in sewage sludge.

Changing the requirements presumably was a political decision made because they could not stand the idea that the sewage sludge had to be disposed of somewhere and given, as in the cemetery, perpetual care. I got into this problem because I was afraid that everybody was going to dump everything in the ocean at one time I keep asking myself, what do the choices involve, whether you put this material on land or in the ocean? I do not think the EPA has yet taken a total holistic look at the possible alternatives. It would seem to me that the situation in California is a classic example of this.

Jorling: No one is going to be able to say, yes, they do. If we had a dictator that could both look at the entire problem and then carry out all the implementing things, I would be scared of that society. That is what you are asking the EPA to do, because if you get into controlling the movement of matter and energy in our society you are in a way the dictator. We have to go through a series of processes that are much more complicated — democracy is messy. Specifically you have been mentioning the RCRA land spreading requirements. Those basic requirements for cadmium are based upon the application of sludge to lands for food crops. But, there are other options. We have a problem with cadmium and certain other heavy metals.

I would disagree with your figures on the efficiencies of pre-treatment in removing cadmium. They are much higher than that in many communities. However, there are still other options. Cadmium should not be used in our society, except in very limited cases, but we have never put it to that onus. We have never asked the question, what should we be using cadmium for? What is essential to our society? The reason we use so much cadmium in our society, is because the Defense

Department (pre-World War I) said that anything that was going to be stored outdoors for twenty-four hours, had to be cadmium plated. That is silly, but it still exists in the Defense procurement regulations, and so every electroplater has cadmium-plating capacities so that they can bid on Defense Department contracts. That is where we can get at cadmium if the Toxic Substance Control Act (TOSCA) ever gets off the ground. You do not just keep saying the choice is air, water, or land. You can go to society and say, why are we using this chemical we are using? How can we avoid using it? Once you are down to the unavoidable use of the chemical, you can make the judgments as to the best way of managing it.

I say again that our ability to know where these materials are coming from, with the exception of a few POTWs, is minimal. When you start hearing all of the requests to ship Boston sludge to sea, and New York metropolitan area sludges to sea, you should be very cautious about what it is you are getting into. I am not asking in a sense for more information before you act. There are occasions in public policy in which you have to act, and, in fact, we do. We are making incremental decisions in the case of Los Angeles and the other POTWs in that area. The incremental decisions will result, I think, in good public policy, and they will be made on a timely, but painful, basis. The California ocean plan requires that the effluent meet certain limits that, with the exception of a couple of the organic pesticides, are measured at the diffusion zone (the others are measured at the outflow pipe). Los Angeles is making progress. We do not have the luxury of saying that here is the solution forever for any of these problems. We are making incremental decisions, and the question is, can we make incremental decisions in the context of a larger framework? Given the fifty-year period, this makes sense.

Steele: I think you are suggesting that we should be in cooperation. What information could be obtained for operation over that period and used as part of the process?

Jorling: We are just beginning to review chemical content of POTW waste streams elsewhere in the country. We can not go into most POTWs and find any capacity to do that sampling, so it has to come from the outside. We are just beginning to secure that essential information; the knowledge of what is in those systems — the nature of it. When we have it, we can begin to make some judgments. Then we can begin to look out there in the receiving waters for effects. That is basic elemental information. I think it is of crucial importance, and we are not even making a dent in that on a national basis, whether it's a coastal outfall or an inland outfall.

Steele: Can you make decisions about what type should be used or do you have that information?

Jorling: Yes you can; I have mentioned the POTW toxic study. The evidence shows that of the twenty plants that have been sampled in detail, all but twenty-nine or thirty of the priority pollutants are in those systems. It shows that the efficiencies of removal of those 100 and some pollutants is over 70% in secondary treatment systems. It shows the primary treatment systems only remove 30%, and these are average figures for heavy metals and organics. That gives you some information. It tells you that you are going to remove a lot more of the priority pollutants in a secondary system than you are in a primary system. That is information. Now, if you can get to the specific component and you find out that it is the result of a single industrial indirect discharger, as it is in many cases across the country, then you can go after that discharger and get some source control.

However, there is a lot of information of a generic type upon which you can make judgments. One of the refreshing things is that the water is getting cleaner in many areas of the country. So we've made some good judgments. They may have been expensive, and they may not have been the most cost-effective, in retrospect, but at least we have made achievement. We have certainly taken the peak off of the pollution loading that would have been expected in growth.

Comment: What are the decision strategies that are used? It seems to me that we need that approach. In effect, we have to make decisions; gather all the information, and make decisions incrementally when the conditions are certain. You analyze the risks of the first step.

It seems to me that approach has to be followed in this balance between land and water disposal. Land disposal of sludge can be watched — the experimenter is aware that the effects can be tested. You know where it is, you can control it, and at the same time you are working on pretreatment. At some point you may have the flexibility to go back to ocean disposal of sludge when the content changes. In other words, it seems to me you are giving yourselves too difficult a problem in trying to treat it as if it can solve all the problems. You can analyze and weigh all the alternatives at once, but what you need to be doing is calculating a step-by-step process to consider the risks of reversing the damage along the way.

Jorling: One of the things that I find very curious is the suggestion that we know less about managed terrestrial ecosystems than we do about the unmanaged ocean systems (in the sense that you can not put boundaries around them). I wonder how widely shared that perception is among biologists?

Bascom: Concerning the question of how widely shared the perceptions are; if we define scientists as university-trained people who are actively engaged in the business of ecology, I really do not think it would be very hard to get a consensus on the things I have said. All I have really done is give you data which we can support. Relative to our studies, we have been to specific places on specific dates and taken data in a specific laboratory and run it this way. We have done lots of it — thousands and thousands of times. I have not given you any opinions on it; I have simply said that this is what we measured out there and what was found. I can not imagine there is going to be anything much more scientific than direct measurements. If there is something else that some of these other scientists would like us to measure, we will go measure that too and tell you what the answer is. We want facts, whatever the outcome may be.

As for the question of land *vs* sea, in this particular example that I gave this afternoon, the sea is a place where we have steady conditions. The sludge discharge has been going on for twenty years, for better or for worse, and at least one thing we know is that the sludge has been restrained in a small area. Certainly all the effects are not good. I have pointed out the bad features of it; the fin erosion, the terrible BOD, and the contamination.

On land, you really do not know any of those things. The idea that somehow it is better to fertilize the land with "night soil" than it is the sea is nonsense. I can think of no rational basis for that. The worst part of all is when you get this toxicant containing sludge in the ground water.

Jorling: If it were night soil, I would not really be concerned whether it went on the land or into the sea. The problem is that it is not any longer just night soil; it is contaminated with a lot of chemicals that have differing characteristics and different ecological roles. I look at a Muskegon, Michigan, and I examine the quality control over the movement of those chemicals in biogeochemical cycles, and I say the problem is a great deal more under control than it is in a Los Angeles outfall.

Bascom: We find some amusement in noting that in the Owens Valley Water, which represents about half of what Los Angeles drinks, the levels of arsenic and copper in the drinking water are higher than we are permitted to discharge at sea. Having drunk it and used it, we now have to clean it up to put it in the ocean. Some people do not think that is so funny.

Jorling: That's a facile response to my question. Are managed systems on land better known and controlled than systems where you are discharging high volumes of waste into the ocean environment?

Comment: The thing that scares almost all of us, particularly here in New England, is the problem of the groundwater. If you could guarantee me that I could find a place where I could put this stuff on land and it would not get in the groundwater. . . .

Jorling: You are giving me Hobson's choice. I want to go back and find that stuff — where it is getting released into the environment and . . .

Comment: (continuing) Until such time as you can tell me what the answer is, I would much prefer to take my chances putting it in the ocean where it will not get in the groundwater, than putting it on land where it will eventually get into the groundwater. Now I agree we have to do some work.

Jorling: You are giving me generalities. If you give me a specific plant in New England, I will come up with a managing scheme which will avoid both of those problems.

Comment: I will take Willard Bascom's point of view that the stuff seems to seep into the groundwater in spite of all of the knowledge that people think they have.

Jorling: Environmental management has moved past that.

Comment: Let us put it in very close perspective; there are six million tons of sewage dumped in New York Bight every year. The highest PCB levels I have seen in the ocean are at the New York Bight dump site for sewage. Where are you going to put six million tons of New York metropolitan sewage next year?

Jorling: Let me just give you some figures and dates. In the New York metropolitan area by 1990 there will be 1.4 million tons per day of sludge.

Comment: (continuing) Where are you going to put it?

Jorling: (continuing) The present plan for 0.8 million tons of that sludge is various forms of combustion, high temperature combustion.

Comment: Does burning sludge get rid of the PCBs? This is an expensive . . .

Jorling: Yes, however, they are not burning it for the purpose of getting rid of the PCBs; that was not the reason the decision was taken. It was the most cost-effective solution to their sludge problem in the five New Jersey metropolitan areas. You are throwing together a whole series of issues. The sludge that is being dumped now in the New York Bight comes from twenty-three different POTWs. Not all twenty-three have the same chemical composition of their sludge; they are not all laden with PCBs; that problem comes from two of them. That can be addressed separately: if we can disaggregate, find a chemical composition, and then address that, we can come up with environmentally sound ways of doing it. The more we disaggregate the waste stream, the better off we are.

Comment: We have effectively cut PCBs off from the source, but the problem exists just the same.

Jorling: We have cut them off at the source in the sense that they are no longer being manufactured, but there are millions of tons still in society. They are in every electronic gadget that we are accustomed to working with. The EPA regulation on PCB disposal is incineration in high temperature.

Comment: Mr. Jorling, it is unfair to compare a managed terrestrial ecosystem and ocean disposal.

Jorling: That is just what Willard Bascom said. I was asking him if that was a shared perception.

Comment: The unfair part of it is that you simply do not manage underground water.

Jorling: I can manage the surface ecosystem with the administration of chemicals to avoid that problem.

Comment: If you are avoiding the problem, then you are avoiding the problem of land disposal.

Jorling: No, I am not. The materials in the Muskegon system are not reaching the groundwater. I do not want any materials to get in the groundwater either, but I am disturbed by the perception of an environmental management in which you have to choose which of your environmental systems you are going to contaminate. I think that is a failure for industrial society.

Comment: You do not contaminate the oceans by the discharge of some materials. I think we can introduce some materials to the ocean without causing an undesirable change in contamination. Of course, the oceans are not panaceas for all waste. We are ending up with a lot of arsenic on land as a result of mining and industrial activity. Where do you put it? It has to go somewhere. The oceans can accommodate a lot of arsenic if it is placed in the appropriate site.

Jorling: If we could get pollutants in nice, discrete forms then certainly we could manage them even more effectively. However, oftentimes they are not discrete, and that is why I say the more we disaggregate the waste, the more we get it into its specific components, the more management options we have available to us. The concern that I hear is that everybody is willing to aggregate wastes and then just choose which ambient environment we throw them into. I think that is a failure on the part of the scientific community and on the part of the legal and policy community. It is unnecessary and, over long terms of time, is destabilizing.

Schneider: I would like to share an observation and it really isn't mine, but it was made by Kenneth Kamlet from the National Wildlife Federation who gave a lecture in a course I was teaching this semester on multimedia waste management. Mr. Kamlet is amazed at the audacity of marine scientists who sit and make decisions about ocean dumping of various materials when there has not been a very thorough study of the other options.

There is a need to take the various types of wastes that we are dealing with and examine the various options for getting rid of them. We need to look at the types of things that Mr. Jorling is talking about. If, for example, cadmium is the real problem, then where is it coming from, can we make alternative strategies to deal with it? Then second-arily, we have to look at realistic costs; we have to try to assess costs to the environment and also try to give fair costs to things when we recycle them. If we have an ability to save a billion gallons of water in Southern California by recycling within the system, or to take the nu-trient-rich material that we have and recycle that back into society, we should attempt it.

I think EPA and other peoples involved in this area, should sit down and try to get at the various options. At this conference we have many people who know a lot about marine systems, and although some of us have worked in terrestrial systems a little, we really do not have experts here on terrestrial systems. Perhaps a very exciting and impor-tant thing for our society to do would be to try to formulate such a task that would really look at the types of alternatives that we have for these problems in the future.

UNIT FIVE

Future Prospects and Strategies

Photo courtesy of Willard Bascom

IN MY VIEW THIS CONFERENCE ADDRESSES ONE OF THE MAJOR ISSUES OF OUR time, the impact of marine pollution on society, and there are no easy answers.

There are also a few simple and fundamental facts we cannot ignore. One is that the world population is four billion people and growing. A second is that the gross national product, on a worldwide basis, is growing faster than the world's population, and in some parts of the world it is growing *considerably* faster.

The third simple fact is the second law of thermodynamics, sometimes referred to as the "no free lunch law." There is no way that one can transport chemical elements from one place to another, combine them in various ways, distribute the products in the ways necessary to drive our society, then redistribute them in their original, natural form and location without significant cost and energy. There are no perpetual motion machines. Once you accept the concept of the "no free lunch law," then you accept a certain level of pollution. From that point on we are only arguing about the details.

Listening to some of my conservationist friends I am sometimes of the opinion that they think they can repeal the second law of thermodynamics, just as the patent office occasionally still gets claims for perpetual motion machines.

On the other hand, there is no reason why we have to continue to accept a linear relationship between the GNP and pollution. We can, indeed, approach the perpetual motion machine, although the details are awesomely complex, and there are no obvious best solutions — at least not yet.

The Center for Ocean Management Studies was established three years ago at The University of Rhode Island to look at some of the tougher marine policy problems that this country faces — and the world — and we try to do this in various ways. Often we bring together a mix of people of different disciplines and backgrounds; it is not always easy as the following remarks, which were heard during the course of the conference, indicate.

Some of my natural science colleagues have been criticized as applying the "count the number of angels on the head of a pin" approach to marine pollution, and by others as measuring what we can measure, rather than thinking about what we should be measuring.

On the other hand, some of the social science and legal participants seem to have their own special law, which can be simply translated as the softer the evidence, the harder the conclusion. Or as I heard someone say, "Haven't these guys ever heard of a standard error of estimate?"

A third conversational trend has been that somehow none of the natural or social science studies have any impact upon those people who make the legislative and political decisions. All this effort is irrelevant.

I believe these comments are overdrawn, but I also believe there is sufficient truth in all of them that we ignore them at our peril.

In this section, we are looking into the future, a particularly important subject to me. I am presently serving as a member of NACOA — National Advisory Committee on the Oceans and the Atmosphere — which is charged with providing advice to the president and to Congress. We have a NACOA panel, which I am chairing, that is looking at marine pollution, more specifically, waste management in our oceans and estuaries. Our hope is to have something useful to say on these matters by sometime this fall. As chairman of that panel I would like to share my perspective on the issue that the interest in marine pollution has peaked and that there is less interest in marine pollution today than there used to be.

I agree that there has been a change. The seventies was the time for legislation. We passed NEPA, the Clean Air Act, the Federal Water Pollution Control Act, the Coastal Zone Management Act, the Marine Pollution Research and Sanctuary Act, etc.

With the passage of the Resource Conservation and Recovery Act in 1976, and perhaps more importantly with the beginning of the dissemination of the Federal regulations with respect to that law, which is now taking place, we are approaching what is essentially a complete set of waste management regulations. We are about to come face to face with the "no free lunch" problem. The ocean pollution laws were among the first to pass and to be implemented. Interest in this country now centers on such things as Love Canâl, acid rains, polluted ground waters, and with this new interest comes a new look at the oceans. What alternatives do we have for the PCB-laden dredge spoils in New Bedford Harbor, or for the wastes from a hundred or so relatively small metal plating plants in Rhode Island; or what do we do about the sewage sludge of New York and Los Angeles?

I disagree with those who say that there is a decreasing interest and concern about marine pollution, but I am convinced that the legislative orgies of the seventies are over. It's going to be a much tougher ballgame in the future as this country grapples with the problems of what is the least harmful solution for a given waste disposal problem. Those who are concerned about marine pollution are going to have to learn to understand one another — engineers, lawyers, scientists, economists, policy makers and academics — which is one of the reasons why COMS holds meetings such as these. We in NACOA need your help and advice.

John Knauss
University of Rhode Island
United States

CHAPTER 17

Strategies for Marine Pollution Research

JOHN H. STEELE
Woods Hole Oceanographic Institution
United States

The title of this book, "Impact of Marine Pollution on Society," suggests that on one hand you have marine pollution and on the other you have society. But any definition of pollution has to bring in the societal component. One such definition is from a book called *Politics of Pollution* by Davies and Davies. They make the point that the definition is dependent on the public's decision as to what use it wants to make of the environment. Only by linking scientific knowledge to the concept of the public interest can one arrive at a working definition of pollution.

The public interest is strong in regards to the ocean, particularly because the oceans are not only a matter of commerce, but they are probably one of the major sources of our mythology, and have always been a source of imagery for birth and death. We have the painting of Venus rising out of the ocean on a halfshell; we have Vikings being put out to sea on a burning ship for burial. I'm sure that the former would cause consternation in FDA regulation, and the latter subject would probably be completely prohibited by the EPA nowadays.

I think one has to be conscious of the fact that the sea has this unconscious symbolism for us all and that the emotions that are aroused are not peculiar to any group and apply as much to those concerned with basic science as to those concerned with commercial use, or to those who look to the sea for relief from the pressures of terrestial life. This underlying aspect is bound to affect our views, decisions, and concepts of public interest.

The three case histories discussed previously present very good evidence that can be used. But the question is how far can the evidence be used beyond the particular case to which they refer and influence future decisions? The question that arises then is whether these case histories are merely anecdotal. They are of interest, but they are not hard although they may, in a sense, form a basis, as in the legalistic

280 Future Prospects and Strategies

sense of precedence. How far can they be used in the scientific context as a basis for generalization? How far can we take these examples and generalize for future cases, or future problems?

One of the things that I think has emerged from the North Sea in regards to marine pollution is that at the highest public interest levels (i.e., the levels of the commercial fisheries) it is not possible to see any evidence of the effects of pollution. As Dr. Waldichuk stated "Clearcut evidence of sublethal or even acute effects of pollution on stocks of strictly marine species of fish is difficult to find."

Certainly, one of the problems in the North Sea was the lack of actual evidence. For this reason, one of my former colleagues, Bob Johnson, carried out a series of hypothetical calculations, starting from the idea of a tanker accident in the open sea, spilling four hundred thousand tons of oil at a time when it would have a major effect on fish larval development, such that the time and the place would be the worst possible case. The results of the calculations estimated a monetary loss at five million dollars. While this is not an insignificant sum of money, the total yearly value of North Sea fisheries is about a thousand million dollars. It was interesting to see the results of surveys for the *Amoco Cadiz* estimating the cost to the open sea fisheries to be five million dollars. This is probably too good a coincidence, but it does suggest that the events, while not exactly similar, are of a certain order of magnitude whether constructed theoretically, hypothetically, or in practice.

A second conclusion concerning the North Sea was made quite effectively by Ian Nesbitt, who is with the Massachusetts Audobon Society, who has suggested that the major lesson to be learned is that the most serious impacts are not in the sea, but ashore. They are the effects, beneficial or deleterious, on local populations in terms of their social and economic structures. The same conclusion might be reached on the *Amoco Cadiz* incident. There may be a certain pattern that can be derived from the different aspects of the experiences which should be considered for application to other situations in the future. But the reasons for different views and actions in different situations will emerge from the different public interests as defined by the sociology of a different area, rather than from the different ecology. We need to be very conscious of that fact when we are considering the relationship between scientific information and the matters of public interest that arise in these problems.

The public interest should be a major part of the decision-making process. However, it shouldn't be completely dominant to the use of scientific knowledge or information.

It would be wrong to claim any high degree of certainty for our general ecological principles, but it would also be wrong to pretend that we just don't have any knowledge at all. One of the things that has emerged — and this is not merely in marine, but also in terrestrial systems — is that the response of systems to changing conditions seems to follow particular patterns that are independent of the nature of these changes, whether they are natural ones, like climatic changes, or man-made ones, like contamination and pollution. We find that systems seem to be able to absorb change to a certain point and then they suddenly respond to it. But that response is not always the death of the systems. Instead, it tends to flip over to another pattern or structure, a structure which is in its own way responding to the nature of the environmental stresses and perturbations. These flips apparently occur as a result of all kinds of changes, both natural and man-made. It would be wrong to put adherent values on the alternatives. One can certainly put values on them, but it is something that we do; we can do it on an economic basis in terms of fisheries, we can do it on an aesthetic basis, and so on.

We can see these patterns emerging and we can use them as general guides. But certainly, one of the questions is whether they are sufficient to provide the kinds of information that we need for decisions. This is where we come to the questions of assimilative capacity and Ed Goldberg's idea of titration. We need to be able to measure the change and the point at which the system flips, and to do that in such a way that we can end up on whichever side of it we choose, or with whichever pattern we choose. We can have some knowledge of the kinds of changes we expect, but we don't have the knowledge to say exactly when it will come.

Ed Goldberg quoted the example of Windscale as, I think, almost the paradigm of this approach, particularly with the evolution of the final discharge level. It is important to remember that when Windscale started, it wasn't certain what the critical pollutant pathways were. There were several that were potentially identified and these were studied. Therefore, the initial discharges were very low, what we consider now to be very conservative. What was done was a very gradual increase based on the evidence. This is an example of the use of an adaptive approach to the management of this kind of system. It allows one, in fact, to concentrate much more on the question of whether the ICRP recommendations are effective.

The views of society are certainly a significant part of the definition of what we mean by pollution and these views are changing. This is something that has emerged very much in the discussions.

There is an emerging opportunity to be less absolutist and more empirical in one's approach, and that perhaps is a personal evaluation outside the purely scientific context.

We are going to enter a period where there will be much greater requirement for quantitative judgments. Therefore, when we consider perhaps to some extent that public interest is decreasing, the requirement for scientific judgment, scientific knowledge, is going to increase significantly.

The case histories that we have considered here give two kinds of evidence. They can be used to generalize in a scientific sense, but one has to realize the need to accept some degrees of uncertainty, and correspondingly the need to allow the available options to be gradually selected on the basis of accumulating data.

The problems can be defined on a very large scale (i.e., national level) dividing them into the elements, water, air, and earth. A different approach may be adopted by looking at these problems in a more holistic way; looking across divisions between the three elements, the three methods of disposal, the three different parts of the environment. It seems likely that perhaps the latter may be more useful as an approach. It is perhaps regional authorities who seem in the best position to conduct these kinds of adaptive schemes.

The attitudes of society towards pollution are changing rapidly as we assess the costs of pollution control in a period of apparently declining affluence. The earlier calls for zero discharge are being questioned both technically and socially. A *Science* editorial was headed "Zero — what does it mean?" There is a view that federal authorities will propose less stringent regulations based on "preliminary" findings that the ocean can assimilate more than previously believed. An interesting question is whether scientific discovery or economic pressure is the major determining factor in this revaluation. In any event, it is apparent that qualitative judgments are no longer adequate.

We are entering a period when there will be greater requirements for quantitative judgments. The translation of pollution problems from some separate, fundamentalist role into the more empirical ambit of a cost-benefit process can have two significant and related effects. The apparent decrease in status could lead to lower levels of effort, whereas decisions about "assimilative capacity" will require a greater knowledge than a "yes/no" approach. *A priori* decisions, often set in a legal framework, will need to be replaced by adaptive management where options are gradually selected on the basis of data accumulated in part from operational experience. These changes are likely to have significant effects on the conduct of research in this field.

The options extend across the land, sea and air boundaries and so we shall need closer connections between traditionally separate areas of research. We shall require more extensive and more critical analysis of ongoing or new schemes for disposal. We should make the greatest possible use of large-scale controlled experiments to provide assessments before the event; and we must learn to use theoretical studies as a component of a total evaluation.

Yet improved observation, experiment and theory cannot remove the uncertainties associated with accidental — or deliberate — perturbations of the environment. Some of these uncertainties are statistical, arising from the great variability of the system in space and in time. We can hope to decrease but never eliminate these uncertainties, and we should be able to quantify them. More significant, and much more difficult to assess, are the uncertainties in our knowledge of the underlying principles which determine the dynamics of the physical, chemical, and biological systems that we may perturb. It is equally wrong to claim a high degree of reliability for ecological principles, or to accept that our knowledge is inadequate as a basis for any alteration in the environment. The scientific studies of any proposal will need to evolve and develop as part of the process from experimental or theoretical studies through to the field program. Similarly, the policies, in general and for each specific case, will need to have the flexibility to adapt to the increasing base of information. How we establish these complementary patterns in scientific study and decision making is one of the major questions affecting the future use of the oceans.

CHAPTER 18

Economics and Marine Pollution

CLIFFORD S. RUSSELL
Resources for the Future
United States

I have found the discussion of marine pollution provocative, stimulating, and in some ways a bit depressing. Let me try to continue the provocation by describing what I found stimulating and depressing. I am stimulated particularly because I have sensed that there are now openings for fruitful dialogue between economists and marine scientists. I am depressed because I caught behind some of the arguments over the state of scientific knowledge, the interpretation of scientific data, and about the public interest and politics, a note of longing for the old Holy Grail of a scientifically best answer — unambiguous, sure to be accepted by all right-thinking people once communicated to them by selflessly motivated scientists, via nondistorting media.

Said more simply and less provocatively, claims have been made that there is something identifiable called "the public interest," and that it is always served by the "best scientific answer." I think that that approach is due to be abandoned — overdue, perhaps — because it is so misleading.

Let me summarize briefly three problems with the "Science will tell us" approach and then let me suggest some of the elements of a more productive, basic philosophy.

First, and most simply, the likelihood of a best answer, even in solely scientific terms, must be seen as very small. The argument about disposal alternatives for sewage sludge should have made this abundantly clear. Even if you find an answer that is scientifically best in one dimension it is likely to be less than best in another dimension. (For example, land disposal of sludge saves the marine environment but threatens to pollute ground water.) In other words, when looking at all the implications — still confining ourselves to scientific questions only — of our policy options, we find that we have vectors to rank; an exercise that only in special circumstances or under strong assumptions that allow us to combine the vector elements into one scalar per policy, will lead to a determinate "best" vector answer.

If we broaden our criteria beyond the "scientific" the problem becomes even more difficult. Even if there were an answer to a policy problem, such as marine pollution by sewage sludge, that scientists could agree on, it would still benefit some people and harm other people. And unless the winners in that game actually compensate the losers, the losers will oppose the solution however good it may be scientifically. That is what politics is all about, and politics should not be a dirty word, because sometimes we are all losers.

The third thing I think is wrong with the "Grail approach" is that some answers that are appealing to scientists — and this is something that Jorling pointed out, and as he is an assistant administrator at EPA, I think we can take him quite seriously — involve unprecedented levels of government interference with society and the economy. An example which comes to mind is the current debate within the EPA about how to allocate among producers and users what the agency conceives to be the annual allowed production of the propellant for aerosols, that we fear is stripping ozone from the stratosphere. Deciding that x tons can be tolerated is all very well, but the question then is how you translate that into action in the marketplace. One possibility that EPA must consider is choosing the firms that will get to use this ration and specifying how much each will get. Well, now, that sounds like a quite different kind of economy than the market variety we like to think we have. And if there are no alternative ways to translate an upper limit on fluorocarbon use into action, we might decide that the cost in terms of moving toward a centrally directed economy would be higher than the costs in terms of larger numbers of skin cancers of settling for some fuzzier solution.

All our decisions will be made under greater or lesser uncertainty. We can't wait for certainty. Even doing what we are still doing, continuing the status quo, is a decision, and that gets made no matter what, if we don't make another. Thus, when I say that we can't wait for certainty, I mean that something will happen in any event.

There are, moreover, well-developed ways of making decisions with uncertainty, and we do not have to choose between two horns of a dilemma. That is, we do not need either to spend an infinite amount to avoid our worst fears, or to succumb to complete inaction and accept the status quo. Rather, we do have to think of assigning probabilities to outcomes and making as rational decisions as we can in the context of our uncertainty (and that includes decisions about what information to get next). Uncertainty is a fact of life and we should not let it paralyze or dismay us. We should accept it and work with it in ways that are fairly well developed.

One part of uncertainty that is perhaps somewhat different and

should be treated with even more respect is the problem of irreversibility — when we can identify it — and it seems to me that attempting
to identify irreversibility is a very useful enterprise. We should be very
careful about trading off short-run gains for irreversible effects.

Now, let me be more specific. The things that I am really supposed
to be talking about include benefits and costs and incentive systems.

First about benefits and costs — I think that we are going to be
able to do better in the future in the measurement of benefits. That is,
as economists, I think we are going to be able to do better, and I assume
that the other parts of that process, which are the parts that scientists
deal with, will be done better, too.

I do think that for any significant problem — the Los Angeles
sludge disposal problem, for example — we shall almost always be in
the position of offering lower bound estimates of benefits. This is both
because we can never really know what the long-term possible damages
will be, and because there is a kind of benefit that people get, which
we refer to as psychological, though it's not a good word. The idea is
that some people gain a real benefit, a benefit that they would actually
be willing to pay for if pushed, simply from the knowledge that our
environment generally is cleaner today than it was yesterday and may
be cleaner tomorrow that it is today. That kind of benefit is nearly
impossible to get at, and so it seems we shall always have great difficulty
obtaining a complete measure of benefits. We will be in the uncomfortable position of not being able to say that the benefits justify the
costs of a particular problem. Rather we shall only be able to say what
the *unmeasured* benefits would have to be to justify the program.

I do think that this is an area in which the cooperation between
marine science and economics has been fruitful in the past and is most
likely to be fruitful in the future. But let me say something about what
benefit estimation involves to make clear what I have in mind when I
talk about cooperation.

If we want to measure the benefits of a particular environmental
policy, or regulation, we have to have four kinds of translations available. One is from the policy or regulation into a change in discharges,
whether chronic, accidental or intentional. The second translation is of
the change of discharges into a set of effects on the ambient environment. The third is of these effects into specific items valued by human
beings such as human morbidity, human mortality, or the quality of a
sport fishery. And I caution you that even economists realize that what
is a benefit to human beings depends on whatever upsets a significantly
vocal minority, so that the requirement that damages be related to the
values of human beings need not imply acceptance of crassly materi-

alistic values. The fourth translation is from those effects into human damage; that is into dollar terms for comparison with costs.

None of those is an easy step. They are all fraught with difficulty, uncertainty, ignorance. You can see where the different disciplines must be called on for contributions, so I won't belabor that point. But I think that in such efforts, if anywhere, marine science and economics come very close together, and groups of people can work together in tremendously fruitful ways. (cf. The *Amoco Cadiz* case study in this volume.)

On the cost side of that benefit-cost analysis, we are usually in pretty good shape conceptually, because estimating cost doesn't involve the problems of principle that we run into with benefits. (And I emphasize that I passed right by those problems when I devoted one sentence to the translation from scientifically predictable effects to money damages.) For example, one of the things that I have worked on in the past is safe drinking water and the Safe Drinking Water Act. Compare what EPA and the water utilities estimate as the cost of implementing that act, and you will find that they differ by a factor of about two. My rule of thumb, which is only half flippant, is to take the average of the costs proposed by the regulator and the regulatee as an approximation of the true cost. The regulator is always low and the regulatee is always high.

Now, about implementation techniques and economic incentives. The system we use now in most of our pollution control work is commonly called, "Command and Control." That is, we tell people what to do in fairly great detail. We specify treatment processes, and as I said in the case of the aerosol propellant, we may specify even more finely: i.e., who will produce how much, when and where.

I believe that is exactly the wrong way to go because it is focused narrowly on achieving with certainty, by detailed specification, solutions that we cannot be sure are best or even good. Further, the specifications tend to be in a form that shuts out new ideas and imaginative solutions. They essentially freeze our knowledge in its current state.

This is being recognized. It is not an ivory tower idea. Certainly it is not my idea alone. Even EPA has a group of people agitating within the agency for more experiments with different kinds of implementation rules, regulations and systems. In short there is a lot of attention being paid to the shortcomings of command and control and to the possibilities for better systems. I think it is one of those areas in which there is a real chance of change, of good experimental work or of some really good new legislation if everything goes right.

Now, what do I mean by a better system? Well, I think a better system would mean more flexible tools, tools that concentrate on giving everyone involved serious incentive to think about discharges and how to reduce them. You can't win by exhorting people to do better (to pollute less) in these cases because it's not in their self interest to do that; and whatever we may think about ourselves individually, I think it would be folly for us to assume that people at large are going to act contrary to their self-interest. So you have to channel their self interest in the direction of policies or improvements in processes that seem likely to achieve the desired goals in the environment.

What you end up trading off in the broadest terms is some certainty about the present effect — and it's only a relative certainty, for our command and control processes, as you know, do not produce the intended effect in all cases — for a chance at a better long-run solution.

The most promising option, in my opinion, now on the horizon, and the one that is getting most attention at EPA, is one or another version of something we can call marketable discharge permits. These are, in turn, an extension of a current EPA regulation called the Offset Policy.

In any case, the idea is that you decide on some basis — and I will not try and specify what basis — what regional total of pollution emissions you think is acceptable, and then you create permits to emit that much and you sell or give them away to current dischargers in the appropriate amount. Then all discharges to be legal must be "justified" by ownership of a corresponding number of rights. In any case, the existence of these rights, which, if you are really limiting discharges, will have positive prices, focuses the attention of dischargers on possibilities for decreasing their discharges, because such reductions free up rights for sale; the rights become real assets.

In the marine area, for example, one could imagine deciding how much dredging spoil Long Island Sound could take in a decade, and auctioning off the permits to dump that much, perhaps subject to other regulations such as capping.

More generally, we should be looking for better incentive systems in many areas. For an example outside of this conference, consider salvage. It sounds to me from the story of the *Amoco Cadiz* as though the existing incentive system in salvage is exactly wrong. It grew up in the period when loss of cargo was a private affair instead of a public one. We need to think about changing that. Or again, consider spills and groundings. We have to make the cost apparent to the people who create those accidents, and no excuses. Another example is dredging — not dredge spoils, but dredging. We have to make the cost, the full cost of dredging apparent to the dredgees, if you will, those people who want that channel deepened. Right now it's part of the pork barrel and they don't pay. Of course they want a new channel. How much of a channel would they want if they had to pay "x" dollars a mile per foot? Very much less, I suspect.

We spend a lot of effort in this market economy insulating people from the costs of their decisions. We should rather try to bring the cost of those decisions home to them, and then we would find that decisions were made in a quite different way, not necessarily automatically best, but quite different, and in many cases better.

Finally, a note on public interest in politics. The participants in this conference have tended to use "political" as a slur, and I think that that's an error, for politics is what holds society together.

What is usually meant when they said "politics" was something like the pork barrel, that is, an aberration, in a sense, of the general political system, in which some narrow interest group gets the benefits because of the set up in Congress and committees, or whatever. The speakers did not really mean that they think political decisions are bad decisions, they just mean that politics sometimes produces an effect with which they disagree. Of course it does.

It is also unreasonable, I think, to expect that the system is going to change by recognizing and reacting to pure truth and beauty. For example, another task force report is unlikely to make any difference at all in marine pollution. I believe that if you as individuals want to see different outcomes, if you think that the outcomes of the political process are not desirable, you would do better to spend your time, not on another task force, but working for a public interest group, volunteering your expert testimony to them. It could be a local group, or it could be a national group like the Environmental Defense Fund or the Natural Resources Defense Council. Those groups — and this is my final provocation to you — do more in a year to change what happens in marine pollution, than the National Academy or all the other task forces in a decade. If you want to affect something, that is the way to affect it, not by sitting on committees and writing compromise reports.

CHAPTER 19

Commentary
Policy Making and Marine Pollution

JAMES WALSH
National Oceanic and Atmospheric Administration
United States

I would like to talk broadly about marine pollution and how it fits into the context of public policy-making. My experience in the area of pollution has led me to the conclusion that too often we talk about pollution in the context of a Congressional hearing after a tanker accident or a severe environmental incident, in a lawsuit, or in some context where emotions tend to run very high and issues get clouded with ideology, philosophy, emotionalism, or a desire to be on television. I remember some of the stories about the *Argo Merchant* and the *Amoco Cadiz*, describing the behavior — and I can say this because I was working in the Congress at the time — of various federal agencies during these incidents. The image you get is that there is a single television camera and at least fifty representatives of at least fifty agencies, each trying to attract the camera operator's attention. As a matter of fact, that kind of behavior makes some sense, because if you didn't get your face before the camera, sometimes you didn't get money to carry out the programs related to the effort.

I recognize something quite obvious which is that we are in a process of change regarding public policy on marine pollution. A recent newspaper report of this meeting indicated that the environmental movement is now passe, that it is no longer viable. To me, that is a typical newspaper comment because superficially that does seem to be the case. But as some more thoughtful newspaper people have indicated, things are not always what they seem. For example, the facts show that Democrats always spend more money on military programs than Republicans, although the impression is the opposite. Everything is a little bit different when the reality behind the perception is examined. And the "perception versus reality" dichotomy in public policy is very important. I think this dicohtomy explains a lot of things about why politics is viewed, on the one hand, in the pejorative sense and, on the other, as the key method for solving national problems.

In Washington, or in any political forum, people try to manage how they come across, how people perceive them in their jobs. Quite often, that perception is unrelated to reality, simply because our national problems are so complex and because most people are unwilling or unable to focus on those problems and form a thoughtful as opposed to superficial view of them. So we all seem to skate along on the thin surface of perception that causes politicians to be politicians. A politician's goal is to create the perception among the electorate that he or she is attacking national problems. Failure to achieve this perception means a loss in November. The true effectiveness of a politician may never get a thorough analysis in this process.

Right now, I think the perception of environmental policy is changing to a sense that it is no longer important. I do not believe, however, that reality mirrors this perception and that our environmental programs are waning. It's the same with the so-called war on poverty. Where has the war on poverty gone? It started in 1965 with 800 million dollars and it's being waged today with 20 billion dollars. We don't know whether we're winning it or losing it, but it is still being fought. This is also true with environmental programs.

In the last ten years an enormous amount of government process has been established to take into account environmental concerns, and they are routine today. I don't think these procedures will go away in the future. We are now mindful of the environmental consequences of the decisions that we must make. As a matter of fact, we are probably more mindful of such consequences than other considerations. We have a very strong environmental program in this country. And I believe that, at base, the American public holds strong environmental concerns, whether it's marine pollution or concerns about lifestyle and the influence of the environment on that lifestyle. So I would have to disagree with those who say that environmental movement, which was born in the sixties and the seventies, has died. It has simply become part of the establishment.

What is changing, however, I think, is the perception of environmental concerns. It is changing because of a number of things. First of all, we are paying attention in the newspapers to other concerns, to energy and to economics. It's like Will Rogers said "Everything I know is what I read in the paper;" and if you don't read about environmental concerns in the paper, you don't think they still exist — but they do. The perception is changing. Another change that is real, however, and has to do with a key element of public policy, is the way we approach uncertainty in environmental issues.

Politicians are hearing more and more from their electorate about jobs, inflation, and the state of the economy. "It's nice to have a clean environment, but if I haven't got a job, I'll wait a little longer for a clean environment," is what some are now saying.

Politicians believe they will be rewarded, therefore, for doing things that will help alleviate economic problems or deal with energy. That doesn't mean to say that the electorate has rejected environmental concerns, but that right now economics and energy are on the front burner.

Changes in the way courts approach environmental problems are also evident. The best example, I think, is the decision of the court in Alaska concerning the Beaufort Sea, where offshore oil drilling is likely to occur and where the most endangered whale — the Bowhead — is found. The court reviewed the actions required of the Federal agencies and recognized that a risk existed. But the Court decided not to stop the lease sale simply because of that risk and said that matters were not at an "irreversible" stage and could be corrected later. Courts a few years ago would be reluctant to allow such action prior to a better definition of the risk. I believe that judges who read the newspapers are good signs of changes in attitudes in society, in politics and in public policy.

We are beginning to see a real change in this area of policy. We are beginning to see that uncertainty is no longer a reason to slow something up, or not do something, that might affect the environment.

However, government policy is still being guided by statutes that were created in the sixties and seventies. As an example, the basic philosophy of the Clean Water Act is that if environment impact is uncertain, or cannot be shown that it will not have an adverse impact, then dumping of pollutants in our waters is not allowed. In lawyers' terms, the burden of proof of demonstrating no harm has been placed on the person who wishes to carry on the polluting activity. When those laws were enacted, the prevailing view held that the market system inevitably placed profits ahead of public interest, and that what was required to protect the public's interest in a clean environment was a strong government program, based on laws to protect that public interest.

I predict that in the next several years, pressure will be brought to change the manner in which we deal with uncertainty under our statutory program. I think more and more you will see provisions that seek to balance economic and environmental interests. Some of this will be as the result of backlash; some will occur because of success in protecting the environment.

I think statutory changes may begin soon. We have already seen a major change in the United States Senate with the departure of Senator Muskie, who was very much an advocate of the "if you are not sure what will happen, don't do it" attitude. That attitude is shared by Tom Jorling, who participated in the drafting of several environmental statutes for Senator Muskie.

In sum, the new front-burner political concerns of economics and energy are fostering change. But again, I don't think anyone should interpret that change as a wholesale discarding of environmental concerns and programs.

There is another reason why I believe change will be coming about. During the last twenty years, our society embarked upon a social experiment to see whether government agencies could define and carry out the public interest without having more losers than winners in that process. I get the sense that the public feels we have failed in that experiment. As a matter of fact, the government has come to be perceived as an entity that has not solved our problems and may be a major source of them.

Right now government agencies are fighting a proposal in the Congress to change the burden of proof with regard to agency actions. Under existing law, when a court reviews an agency regulation concerning e.g. the dumping of dredge spoil, the judge follows an automatic presumption that if proper procedures are followed, the agency made the right decision and the court will not substitute its opinion for the agency. The burden is on the challenger to provide that the rule is wrong and must carry that burden before the court will overturn the agency action. The amendment in the Congress would change this presumption, and would require a Federal agency to prove to the court's satisfaction that it did the right thing and that it resolved all uncertainties before a rule is promulgated. The burden is therefore shifted to the agency.

In public policy terms, such a shift would represent radical change. It could mean that agencies would not issue regulations whenever uncertainty is involved. This would shift pressure debate to understand the impact of pollution from the polluter to the government. Whether the amendment will be accepted or not is open to question. I would say that only a minority of people feel that that is an acceptable change in public policy at this time.

Nonetheless, it shows that as a society we are beginning to doubt whether government agencies and the regulatory process are the best way to protect the public interest. I don't know where we go from here. Initially this country was very dependent on the private sector. Then we saw the growth of government in the last few years, and the courts have gotten more heavily involved. If we reject government agencies, do we go back to relying on the private sector?

I do not believe that the fights we are having over the environment and marine pollution are likely to go away. A fight over a polluting activity is not necessarily always one involving environmentalists *vs* developmentalists — a public interest do-gooder on the one hand and the corporate entity on the other. Quite often you discover that it involves a competition among economic interests. Take the Portsmouth Refinery case, for example. The fight there was between a fishing industry that felt itself threatened by a refinery. The refinery operators believe, however, that the transportation of oil is much more important than the local activity of shellfish and finfish harvesting. There are other examples.

Harper's Magazine pursued a very iconoclastic editorial policy against environmentalists as being nothing more than rich people who were protecting their private property from encroachments by various types of development such as the Storm King power plant. One could probably make a good argument that our basic motivation in environmental issues is basically economic.

Thus, I believe that debates over the environment policy will continue and concern will remain high. Our perceptions of the issues may change, but the reality of trying to protect the environment for a variety of interests is well established and will not abate.

Let me turn to another point. I believe that we are past the period of major wholesale legislative change. I have come to believe that the idea of some all-encompassing policy apparatus is neither come about nor it even makes a difference. We are really the era of implementing statutes, and carrying out established programs. We have answered the easy questions, and there again I agree with John. I think we are now getting to the tough ones, to the questions we know less about.

We will thus begin to re-examine regulatory standards such as those contained in the Clean Water Act which speak of preventing "unreasonably degradation of the environment" or assuring continuation of a "balanced indigenous population". If anyone can tell me what those standards mean definitely, I would be very surprised. I don't know what a balanced indigenous population is, and I don't think anyone ever will in the future. Nonetheless, it is a standard in a statute to which we must adhere, unless changed.

What will happen over the next ten years in this era of implementation will be to refine those standards. I think in many cases we will find them unacceptable and revise them. We won't do away with the statute, and we won't do away with the need to have some kind of a standard. However, if trends continue, the standard will probably be reflected in absolute protection and more in balance. It will be reflected in amendments to the provisions of the acts that create the standards by which federal agencies administer their programs. Like it or not, this is the trend.

We will also, I think, see better communication in the era of im-plementation. We are all beginning to realize that specialized disciplines must talk to each other, must learn about other disciplines. We are learning that one should not be afraid to talk about the law if one is a scientist, or to talk about economics if you are a lawyer. We must master the general notions of various disciplines because the interaction of all disciplines is needed to truly comprehend a complex pollution problem and formulate an acceptable answer.

In the marine pollution field, we have not had a history of good communications. Communications — between and among disciplines, those in the private sector who are engaging in activities that may be polluting, and federal agencies which have a variety of responsibilities — is something that will come about in the future and must, if we are, in fact to marshal our resources and be able to cope with the complexity of the pollution problems we now face.

Let me conclude by talking about what I feel is one of the critical things that needs to be done — that is to put together a much more rigorous research program.

I believe that we don't need to put together new, more rigorous programs for deciding public policy, for the simple reason we have such an apparatus involving Congress, the Executive Branch, and the various local governments. Like it or not, this process is not likely to change. You may feel better when Congress is in recess — but they are not going away. The system can be manipulated for good and for evil. I won't tell you which side I am on — you can judge for yourselves. We do have a policy process and institutions that can work very much to the good if rightminded people are in charge. It can also be the tool of irrationality if not watched over by the rest of us.

Therefore I believe we must concentrate on improving our under-standing of the ocean pollution problem. I am not one of those who believes that scientists are always seeking to research their own pet ideas. There is a measure of that in all of us, but it isn't as overriding as the cynics believe.

My impression is that there is so much more that we need to know about the sources, fates, and effects of pollution in the ocean, that we simply must attack the ignorance that pervades this entire area.

I don't believe we are spending enough money on ocean pollution. But, at the same time, I seriously doubt that we will be able to get much more in the short term. Funding seems to be best in times of polluting events and worst when such events disappear from the front pages.

Consequently, we do not do enough advance work to prepare for a disaster or enough follow-on work to learn from one. For example, we did not do enough work, I think, in preparing ourselves for an *Ixtoc* type of accident. And even if such an event occurs, one cannot simply walk across the street to the Office of Management and Budget and get money to respond to it. I have discovered that the process is just not that flexible.

Reasons wholly unrelated to solving marine pollution problems — such as balancing the budget — sometimes prevent us from doing what we wish to do.

We should focus more intensely on pollution events. We should gather the best experts available to address a particular problem, whether it's municipal waste or radioactive waste disposal and then to communicate what they know in a very articulate fashion to decision-makers.

One of the biggest problems in our government is ensuring that decision-makers have the best information possible before they make a decision. Unfortunately, in this regard, I have found that good information might get published in a professional journal, but that some information is not made available on a timely basis for decision-makers. I realize the problem is complex, but we still have not developed techniques to explain decision processes to scientists and to explain science to decision-makers. We need to develop new methods of communication that work better than the ones we now employ. This conference is one part of that effort.

Also, I believe we need to develop throughout the country centers of research excellence that focus on local pollution issues. Did you ever notice how many centers of energy research there are growing in the country? How many centers for marine pollution research exist in this country? We do need centers of research — in the private sector, or at a university, or both — where people concentrate on marine pollution problems, the chronic, the intentional and the accidental. Such centers should have a relatively steady funding base that doesn't go up and down depending on whether an *Argo Merchant* or a *Torrey Canyon* went on the rocks, because such incidents may not be the real problem. It may be PCBs. Such a center should focus on determining the real problems, modeling the environment, or trying to understand synergistic effects.

How we are going to be able to do this, I don't know. I have learned that achieving this kind of thing in the Federal government would be very difficult. The government, and particularly the Office of Management and Budget, believes that research funds, particularly applied research funds, ought to go to agencies that have a functional or mission responsibility. I strongly disagree with that approach, not only because NOAA is not so organized but also because it's hard for a regulatory agency to also be a good research agency. EPA's research program has been criticized, as well as the Department of Energy's.

I feel there are some real problems with a strictly "functional" approach. It's pretty hard to suddenly generate expertise in a particular area, whether it's CO_2 or PCBs in the marine environment, simply because it's your mission to deal with energy or environmental regulations. It's better to use an agency with expertise. But I have not convinced OMB of that view. Unfortunately, the way our government is organized has a lot to do with the kinds of programs we can put together and how well we carry them out.

The five-year pollution research plan surprised everyone because of the size of the Federal effort — a thousand projects and $165 million in 1979. It was most interesting to ask each agency to spell out what they were doing with their money, what their plans are and how did they fit it into their missions. We got back everything from one to fifty pages. Some of the plans were well thought through; some demonstrated that certain agencies were simply spending money in an interesting field.

I would conclude by agreeing with everything that everyone else has said. I personally am committed to improving the methods by which we understand the sources, fates and effects of marine pollutants. Knowing more will help us make better decisions. Without adequate information, decisions will be made on the basis of emotionalism of political ideology, whether it is Santa Barbara or George's Bank. We should bring reason to the debate by eliminating uncertainty to the full extent we can.

CHAPTER 20

Commentary
Discussion

Goldberg: You propose, Mr. Walsh, the establishment of centers of excellence to consider marine pollution problems. With the limited funds available, I think perhaps it would be an inefficient route today. I would suggest a technique that the Atomic Energy Commission (AEC) and the Office of Naval Research (ONR) had many years ago of supporting a group of academic scientists carrying out research relative to their missions. At this time one of their missions was the regulation of the release of nuclear materials to the environment. They would have a powerful expertise that they could call upon when problems arose. An example of how such a group was used came about when I was a member of this retainer group in the early fifties. A satellite with a nuclear power plant (Snap 9A) was aborted, and the debris came back to our planet. The morning after the disaster, the AEC had assembled ten or fifteen people in Washington to seek out the strategies one should use to assess the resultant danger.

One of my objections to centers of excellence where people of various disciplines are brought together to attack single problems is that you are taking scholars away from their normal academic homes, away from where present day concepts and technologies in their fields are being developed. Usually when a single economist, or biologist, or chemist is brought to a center of excellence of interdisciplinary workers, you have detached them from a terribly important scholarly sustenance.

On the other hand, I would argue that, with respect to marine pollution problems, what we need in the future is a large group of academic, industrial, and government scientists who can meet periodically or, if needed, on an ad hoc basis, and who can call in other experts to consider needed activities in marine pollution.

There is one other activity that is urgently needed if such a group is to be effective. This would be a centralization of the relevant data. I was very much touched at this meeting by two statements that were made. One was by Mr. Roosevelt when he was discussing a PCB problem. He said he had one set of data that said one thing and secondhand information of conflicting data. This is a very disturbing situation. We do need centralized data for these marine pollution problems and a corps of scientists whose credibility can be accepted to a much greater degree than the situation which Mr. Roosevelt implies. There was a similar situation when we were discussing the southern California area. Mr. Jorling had information about one scientific assessment which opposed the other scientific assessment.

In summary, if you want a selection of the best scientists to provide you with a base with which to assess marine pollution problems and to formulate strategies, a reasonable strategy is for a single agency to support a large number of scientists in research relevant to missions of the agency.

Secondly, I think we need a centralization of data relevant to the study of marine pollution problems, which is accessible to the experts, or to any other group of people or individuals who would like to see it.

J. Walsh: Let me take that last point about centralization of data. We concur with that and find that one of the problems has been the fact that information is hard to get. As part of our pollution plan, we put heavy emphasis on developing an information system. We have good data banks on climate and on oceanographic information, and we are trying to start one on pollution. I would, unfortunately, have to say that the House of Representatives blessed us with cutting out that money from the budget — for what reason we are not really sure, other than the fact that it was a new start and they wanted to save money.

Nevertheless, we have considered developing an information system along the lines that we have developed with climate and weather and oceanographic information to be our highest priority. So we agree with that and we will be pursuing that.

In terms of the centers of excellence, the thing that troubled me about your comment is that it seemed to be an attack on the university system itself. However, I will put that aside for the moment. Let me take it from a strictly administrative point of view. First of all, putting scientists on retainer is very hard to do under current administrative regulations and concerns. Also, Dr. Goldberg, I would love to have you on the phone, but sometimes you are in Turkey, and sometimes you are in Spain, and if I have a problem — if I have you on retainer and you are not around, then I am in deep trouble. I may have to go to a center of excellence, where there may be two or three of you.

Goldberg: By retainer I mean giving scientists research support, without peer review.

J. Walsh: Being a non-scientist I would be scared to death to give something. They will say Bud Walsh, the lawyer, gave Ed Goldberg, a scientist, without peer review, this money. It is an interesting idea. I question the practicality of such an arrangement, but I like the concept.

Goldberg: But they already do it today.

J. Walsh: But they are doing less and less of it.

Knauss: I think what Dr. Goldberg was suggesting — if I can try to interpret for him — was that in most mission agencies at present, support for university scientists is for very specific tasks, often for very short periods. In the case of the AEC and ONR, they essentially agreed to support a person or a small group of persons in an organization to do relatively broad-based research in a given area, although that area could be narrowly defined, with that support continuing for three, four, five, or maybe even ten years. In return for that kind of support for research in those areas in which they had a mission, the agencies felt quite free to get on the phone during emergencies, or otherwise, to say, "Now, look, you guys, we have been supporting your research. We need your help, and we need it immediately." Perhaps some of the researchers would be in Turkey or otherwise unavailable at any given time, but if an agency had a network of scientists it was supporting, the agency could always count on getting the help it needed when it needed. There would be a knowledge base and well-trained, knowledgeable scientists who could help. Is that what you are suggesting, Dr. Goldberg?

Goldberg: Absolutely.

J. Walsh: I see some problems with that only because not everyone has enough money to do that without scrutiny. That sort of thing does cause problems with Congress, as you know, since the whole attitude of Congress and the public is, "What are we getting for our science dollar and what are these people doing for their money?" There is an irrationality afoot these days, symbolic of which is the Golden Fleece Award, and there is a distrust of that kind of laissez-faire relationship, which probably worked very well for the AEC and probably could work very well now. However, we have to justify our funding on the basis of what problem are we going to solve for the public. How can we say, "I'm giving ten thousand dollars this year to Ed Goldberg, and he is going to help me solve this problem, and, by gosh, it is related to my mission here." I have to come up with something specific, or people will begin to say I have a friend on the payroll.

People are so skeptical that the government is doing the right thing that you have to go through these contortions. It is reflected in the universities which are over burdened with the paperwork they have to deal with now — justifications, reports. It is something that is hard to turn around.

Knauss: As a university administrator, I can only say Amen to that last point.

Waldichuk: I think the objective of the last transparency in my talk has been met; it has provoked people. Although not everybody agreed with it, some did. We all agree that change is taking place; some of this change may be for the better and some of it may be for the worse. If I were to redraw that last figure (see Chapter 4), the only thing that I would change is the caption on the Y-axis, which I had marked as the "Environmental Concern Index," to the "Political Impact of Environmental Concern Index," because I think that the political impact is the important thing that we must face in the end.

I appreciated the comments that were made by Mr. Walsh, and although I think he said that perceptions have changed and reality has not, I think he confirmed my concerns that we are not getting the support to do those things that we should be doing to understand a little better the effects of marine pollution. We are not much further ahead today in understanding many of the problems of marine pollution than we were in 1970. We have introduced some good new legislation since that time. Some of it was brought about, I am sure, in this country as it was in our country, by agencies and politicians, because it was fashionable at that time to get on the environmental bandwagon and establish pollution control legislation, sometimes with not very much scientific background.

If you bring in legislation that says we must stop dumping wastes and introducing other matter into the ocean and you stick by that with rigid enforcement, then there is no need for scientific information on the effects this has; we merely stop it. But we are not stopping ocean disposal of wastes, and we are not likely to stop it in the future. In fact, we may be increasing ocean disposal. We can only hope to control it better. Unfortunately, we are not really learning enough about the impact of waste disposal in the marine environment and the cause-effect relationships to properly manage ocean disposal practices.

My plea today is that we try to get into the field the kind of top-quality scientific expertise that is needed to solve marine pollution problems. The problems are challenging, and the best scientific people should be encouraged to tackle them. We are finding that in the scientific community those people who really have excellence are sometimes becoming disenchanted with environmental work. They are retrenching into their basic disciplines and are looking at fundamental problems. They are going to the National Science Foundation, or in our country to the National Research Council, and saying, "Look, we have an interesting project proposal here. It may or may not have any relevance

302   Future Prospects and Strategies

to marine pollution, but we would like to study it for its scientific merit." If they can get support, they will move in that direction and not necessarily relate their findings to the real problems of marine pollution.

Twenty years ago, those of us who got into the marine pollution problem were regarded as scientific curiosities, or anomalies in scientific research, by other members of the scientific community. I hope that this sort of thing does not occur again, because there are some real scientific challenges in the marine pollution field and the best scientific talent is needed to pursue them.

What might be needed is a focus on a particular problem or set of problems — drawing on an interdisciplinary approach to investigate them. Our major pollution problems on the west coast of Canada are in the estuaries. You also have many estuaries on this coast. One of your big problems is urban pollution, which occurs wherever there are people. I still maintain that on a long-term basis we do not really know too much about the impact of sewage, whether it is treated or not, on the coastal marine environment. Organic enrichment from sewage discharge into estuaries can have a long-term ecological impact.

The focus has to be on the inshore zone. That is where our living resources are, and where the major impact of land-based pollution is taking place. We have to concentrate our efforts in a given coastal area and pursue a long-term study if we are to obtain meaningful results. Our coastal natural living resources are limited and we must understand the effects of our waste disposal practices if these resources are to be preserved. I hope that we do not allow the curve that I showed to bottom out before we gain this understanding so that we can introduce any needed controls.

I should point out that the environmental movement is not dead; it is just not as effective in mobilizing action as it used to be. Moreover, I do not think that the curve that I showed will ever bottom out at the same level as it was in the early fifties. Nevertheless, I feel that downward trend is there, but it is political impact that should be stressed rather than just the environmental concerns of the public.

Curlin: John Steele mentioned the problems of dealing with uncertainty and Clifford Russell suggested that there were other ways of dealing with incentives rather than the straight regulatory approach that is taken by developing specifications, limitations on activities. James Walsh pointed out that there are changing perceptions and that we are indeed in an implementing phase rather than a legislative phase. Annmarie Walsh was right on target with regard to some institutional problems, and I talked about some of these. The factors that each of us discussed fit into an existing framework of institutions and policies into which we are to some extent inextricably locked. Mr. Walsh observes that over the next several years we will see some modifications, at least in the legislative area. We will also see probably a little more innovation and a little more flexibility in the interpretation on the part of the implementing agencies. However, we are still faced with each of our responses to these individual aspects and perspectives that we have on the total problem of marine pollution. We have to fit this into a kind of fundamental framework.

What none of us addressed is the problem of institutionalizing and accomplishing such suggested things as the centers of excellence or dealing with regulation in a different way, or developing a regional approach of dealing with the problems, as Annmarie Walsh suggests. We are still dealing with existing institutions.

Is there anyone here who is going to contribute any ideas about how we make this system work or what the practicality is of altering the system so that we can make a future system work? I do not agree with Mr. Russell that we have to resort to the public interest groups to force this upon us. That is one option, one that I would choose as a last resort because I am inside the government rather than outside. Is anyone going to suggest any alteration in the system, either practical or unpractical, so that we can start working towards bringing all these other things to balance.

A. Walsh: We are really concerned with the social impacts of water quality. There seems to be some consensus that we are moving from a period of absolute goals to a period of decision-making that requires some balance among the wide range of interests. The scientific evidence that I have heard here is that you do not see dire consequences from using the ocean for waste disposal, and the economic evidence is such that we must begin to balance the goals for economic, social, and political interests. What will be the effect on the water quality in a system of decision-making like that? It depends entirely on what kinds of interests are weighed, how they are weighed, and how well they are expressed.

It seems to me that the long-range social impacts are going to be much more important in the future and that research both on the social side and on the technical side needs to be directed towards these. Because we are going to have increased leisure time in the future, we have clearly increasing concerns for long-term public health problems. Research results on radiation, cancer, and genetics are increasing public interest in long-term public health effects, which used to solicit less attention.

Regarding nutrition, in almost any given year there are two to three books on nutrition on the best seller list. There is suddenly more interest than before in the long-term impacts of food supply and the prices of various types of food. About quality of life concerns, revitalization of coastal cities is going to be absolutely essential in bringing back part of the tax base in the future. Basically, I am making a plea that we realize that the water quality can come out as an important goal to end this complicated social political bargaining process, particularly if we get the kind of cause and effect data, or advice, from the scientific community that can be used to bolster the case.

Russell: I would like to respond to the suggestion that we have to make some effort for the system to work by asserting that the system does work. I sense in this conference and in most conferences about pollution a fairly short-term point of view about getting action, despite a long-term view about damages. If we can not change it tomorrow we think the system does not work. But, in fact, it takes a lot of incremental steps and a lot of effort to change something so important and to impose costs of this magnitude on the private sector and governments of lower jurisdiction, and that is probably how it should be.

Bascom: I am disturbed by the suggestion by some speakers that there is some conflict between science and environmentalism. I feel that I have probably been an active conservationist and environmentalist as long as anybody here. I feel that the purpose of science is to give you some facts from which to operate. If politicians and attorneys do not wish to use the facts, well then that is a different question.

A good deal of effort goes into the solution of non-problems these days. One of the things that I have often said is that metals in the sea are simply a non-problem. I realize that most of the people here today probably will not agree with that, but I expect that ten years from today virtually everybody will agree with that statement.

The scientists are trying to furnish people with data on which they can base sensible decisions. I am very happy with the way the environmental movement has gone in the sense that it has already done great things to clean up America. The EPA had a marvelous first-shot opportunity, but now we are at the position where it is worth considering some of the finer points in its program that ought to be modified somewhat to make it more efficient and more sensible.

My second point has to do with the suggestion that scientists do not agree on the meaning of ecological findings. I do not find that to be true. For example, I have a board of scientists who advise my group several times a year. It consists of people like professors Ralph Mitchell of Harvard, Perry McCartney of Stanford University, John Isaacs of Scripps Institution of Oceanography, and others. Using their advice and suggestions every year we plan a program. Then when we first present our program, we specifically invite to our laboratory all of our most ardent critics. That includes people from all sorts of organizations, such as the Environmental Defense Fund and the League of Women Voters, and we generally have two or three representatives of NOAA and a couple from the EPA as well. These people are invited from the very beginning to criticize both where we are going and what we have done in the past. We rarely publish anything that has not been reconsidered and commented on by these people.

I would say that only on the most minor things do we find real disagreements, mainly because these are simple, straight forward scientific measurements of some facts. I would say that we have rarely, if ever, drawn unscientific conclusions. We never advise anybody about what to do or what not to do about waste disposal — that is not the business we are in. But on the scientific questions that we deal with, we would have very little adverse reaction.

J. Walsh: I agree one hundred percent that policymakers ought to make decisions based on best available facts, which should be presented as accurately as possible. People have different standards of proof as to what is required before they will conclude that $E = mc^2$. It takes time to convince everybody. Between the time that you first propose it and the time that everybody agrees, you have a period during which people have differences of opinion.

Risk analysis, in my opinion, involves both the facts and a subjective decision about whether the risk is worth it. On the one hand you factor in all the information you have available — you get the best and set it forth most clearly. Then a decision-maker assesses the risk and decides if it is worth it for society. It is not strictly a technical analysis; it is not strictly factual.

Everybody might approach that second part of risk assessment differently. It is a subjective element — whether it is worth it to take a risk on Georges Bank with drilling. The Secretary of the Interior has a mandate so that in most situations he will probably say that it is worth the risk. If the decision-making authority is given to some other agency, the result might be different.

There is nothing wrong, in my opinion, with having differences of opinion in risk assessment. I would not be upset if there are people who differ about whether the risk is worth it since that is something that has to be settled in the process. While a scientist may come to you and say he agrees with you on the technical issues, he might conclude something differently, based on those facts. He might subjectively decide that the risk was worth it or not worth it differently than you might.

Schneider: Facts, information, data. Quite frankly, the data base that we operate from is at best fractionalized and is, in many of our eyes, filled with many highly dubious numbers. I think one of the first things we have to provide in this nation is an intercalibrated, coordinated data base.

One of the first things we have to do is sit down and provide a peer reviewed means of getting good information into our data banks that the decision-makers can use. For example, have the levels of PCBs in bluefish ever been intercalibrated between researchers? Quite frankly, we do not have a firm intercalibrated data base from which we can make decisions. I was hoping that the title of this conference would be the "Status of Marine Pollution in the United States," instead of "Impact of Marine Pollution on Society." We do not have a good view of the status. There are some places where there has been outstanding work, a few little pockets of good information. Beyond that, we do not have the synthesized and coordinated data from which we can afford to make the kinds of policy decisions that you the decision-makers make.

J. Walsh: Is that true for discipline?

Schneider: While I can not speak for the other disciplines, I have looked at this discipline carefully. It is full of holes like Swiss cheese.

Knauss: Good chemical analyses are much harder to make than good meteorological observations, if that's your reference. There is a long tradition of making good, credible, standard meteorological observations, and so forth, but the good, high quality analytical data, the chemical data, is a different ballgame.

Steele: I would like to follow up Dr. Curlin's question about organization. It is a paradox of the present situation that, as the more emotional aspects of marine pollution get de-emphasized and we accept a more empirical approach, the need for good research becomes much more important. Yet the money available for that research may become less, particularly for the longer term research. We will still have the kind of crisis funding that occurs in response to occasional dramatic events, but we may lose the studies that bridge the gap between basic research and ad hoc response. I would like to see NOAA have the ability to provide that kind of long-term funding, because the one thing we are going to need is research that lies at the boundary — that is both long-term and relevant. I am very worried that we do not have the appropriate organization to ensure that a suitable system of funding will be available in the future.

Comment: One of the things that has been touched on during the last few days is the need for increased interaction with various disciplines. Each of the three speakers addressed that in different ways this morning. My perspective in working with a number of national environmental groups is very strongly supportive of the need for the mix of scientific, legal, sociological, public policy, and economic views to come together and to do more than just exchange information. They must provoke each other into trying to reach some kind of tentative or preliminary answers to the issues at hand.

Last summer I had the opportunity to participate in the Crystal Mountain Workshop, and I think some fine results were accomplished there; information was produced. However, I think more could have been accomplished if there had been more of that mix of disciplines, and more could have been culled from that. I think the same is true of a gathering like this, but at times it has seemed that case studies were presented for the purpose of presenting case studies. Not enough opportunity was provided for the participants to provoke the case studies to what they mean in terms of addressing the problems.

There is a natural alliance among the scientists, the environmental community, and other disciplines, and I strongly concur with Mr. Russell's suggestion that there is not only the opportunity but also the need for scientists to make themselves available to activist environmentalist groups, who may have different strategies, different approaches, but who can work with the scientific community in trying to reach some common objectives.

How we actually accomplish that interchange of information needs to be addressed. To throw it out is one thing, which I think you did in part, but it is another thing to talk about it realistically. What does it mean in terms of future grant decisions that NOAA makes, the contracts that it lets out? What does it mean for institutions like Scripps or Woods Hole in terms of the active willingness of their scientists to seek out persons in the other disciplines to try to arrive at solutions? I think that is a challenge this kind of conference presents: where do you move from here to focus more on the objective scientific side of things to follow through in some realistic way in dealing with the real world — politics, administrators, industry?

Curlin: I sense that when a group like this talks about scientific information for the decision-maker, many in the audience do not understand how the process works in the government. My background as scientist and lawyer, plus experience in the Congress and the executive branch of government, has enabled me to see how scientific information influences or fails to influence natural resources decisions. Perhaps the OCS decision process is the best example of a recurring decision system that impacts and, in turn, is influenced by topics we are discussing today.

Stepping back into my role as provocateur, let me observe that the National Environmental Policy Act, with all of its splendid benefits that have been derived, has managed to distort the federal R&D system, at least as it applies to resource management. As an example, in the pollution area, the Department of the Interior, according to the CPRDM Report, expends seventy-one percent of the money going into this area. Most of it is aimed at oil and gas, with the objective of producing a good EIS. A "good" EIS is one by definition that can stand up to scrutiny by the courts. The science behind that environmental program is sometimes questionable, and it has to be questioned because we are trying to force it into an artificial time frame, an artificial system for developing scientific information. We are trying to address immediate problems, that is the next lease sale on the five-year schedule. I look back historically and see institutions (i.e., NOAA, EPA) that should have been responding to provide certain vital information since at least the early seventies when it was obvious that the OCS program was going to be accelerated. For some reason, the agencies have not responded to this kind of overall need for developing an environmental base of information for the future. So today the nation is caught in this whipsaw; we address the immediacy, we pour money into quick solutions, and we come up with imperfect answers. We are not really putting together any kind of enduring data and information on which we can base future prospects.

How do we turn that system around? I do not know. It is an institutional problem forced on us by law and imposed by our governmental framework.

J. Walsh: I disagree with Dr. Curlin about reorganization. The proper reorganization that you would decide upon would not take place because most reorganizations generally come down to personalities. Also, I do not think that the system is quite as distorted either. Obviously, in applied agency activities the regulatory process, the NEPA process, and the mission-related process are important. That does not mean that it is bad because it is being spent in that way. You will discover that a lot of the pollution money is channeled toward information-gathering types of activities that are of general benefit for the scientific community in a basic sense. We also have a lot of research going to the university communities that is not related to regulatory deadlines.

I would agree that the balance may not be proper, that we may be putting too much into the short-term such as the environmental impact statement or the mission-related work, for example, of the Army Corps of Engineers. The Army Corps of Engineers has an interesting process; they take the pollution research money and distribute it to district engineers who decide how the funds will be used; usually the allocation of funds to regulatory requirements then in effect.

I would not say that is wasteful because a lot of the research money is used to address those problems that need to be addressed at the time. But, I would say that we do not do enough long-term, fixed, consistent, steady state research funding on various kinds of pollution problems. It is a serious difficulty which is related to the fact that OMB and Congress live two and three years in the future. Beyond that it is somebody else's administration, and as long as it is controllable for this year, we can always put it off.

That kind of attitude can only be changed by just putting together the right kind of people to make a convincing argument for what has got to be done, and by using the pressure points of the political system.

I am not in favor of major reorganization because I am not convinced that you can get rid of all the shortcomings of a reorganization while attaining the benefits. I just have not seen it happen.

Curlin: You did not hear from me anything about major reorganization or reorganization at all. I am talking about making the present institutional framework work. Now, if somebody conceives that as meaning reorganization in the sense of moving boxes, then so be it to them.

Saila: I think I agree with what the three previous speakers have said, but I would say it somewhat differently. I am concerned about the priorities and the direction of future applied research dealing with this problem; I think there is not enough flexibility in the system now. I would take as an example power plants and marine pollution problems related to them. I think there is an order of magnitude. More money is spent on learning about striped bass in the Hudson River than about the disposal of the radionuclides from the nuclear power plants located there, and to me that is disproportionate, in spite of the fact that I am a fishery scientist. To me this is an unfortunate choice of priorities. I really do not know the reason for this. I feel the same way as I look at that blackboard and I see a very small set of alternatives there for sewage treatment and disposal. I perceive other alternatives which are not listed there, and I am not sure why they cannot be considered as well. Mechanisms could be set up to provide more flexibility in considering the alternatives.

Davies: I would like to go back to something Mr. Walsh said that has become very evident in a couple of experiences I have had in the last years working in the Great Lakes, with kepone problems in the James River, and now with Chesapeake Bay. Mr. Walsh says the big problem is we publish in scientific literature, and we spend very little effort putting our analyses together into some sort of synthesis that helps a decision-maker and communicates with the real regulators. Allan Hirsch, an EPA administrator, said it rather well. He said most of us as scientists are at level four, if you like, at an intellectual level and at a problem level, and the people who are day-to-day regulators, day-to-day decision-makers, are down at level one. The tools that they are using are level one tools. I guess I object to the way Dr. Schneider referred to the local regulators. We have a tendency to be academic snobs and look down upon the day-to-day monitoring and rather characteristically tear apart the EPA's computerized data base system as a useless system.

A division is created between the people who are day-to-day regulators, the states, the local general-purpose government, perhaps the EPA, and the academic community when we are saying that we are going to do research on very specific problems, high quality academic research, but we are not going to relate to the people who have the real problems.

I feel that we are very deficient in doing this. For instance, one of the first things we tried to do in the Chesapeake Bay Program was to write a state of the Bay report which included where the Bay has been, where is it now, and what are the stresses on it. If you looked at the scientific data that are available and try to put that together and come up with some sort of systematic evaluation, you fail; the data just are not there. We have not done day-to-day monitoring. We have been doing sophisticated things, but we have not been collecting the DO's and things like that simultaneously and giving synoptic data sets that can be used as data bases for the future. I feel we are very deficient in this: working at level four and not communicating with the people who are actually doing the operating and doing the day-to-day regulating at level one, we really have to watch our reputation.

Comment: I am a free market person, but I am going to say something counter to the idea of incentive. Mr. Russell implied and other people agreed that somehow we can create a better system of controlling discharges if we use economic incentives. Of course, that is an idea that economists have been kicking around the last thirty years or so in the pollution literature and in the economic literature before that. I think there are some basic problems that ought to be brought out when one talks about incentives as a new approach. For such a system of incentives to be truly better than the imperfect system of standards and regulations that we have now, there are some basic things that have to happen, and I do not think they are going to happen.

First we have to be able to quantitatively define some objective function that we want these incentives to meet, and we really do not know how to do that now. That is what standards really are. Standards are computed values, subjectively computed values, of what the basis might be for pollution control. I do not think we can do that any better for incentive creation than we can for standards; in fact, we can not do it for either one. We can also impose on the other hand a set of taxes of charges and they will do the very same thing as incentives in a theoretical situation. However, I just do not see that working.

Secondly, for a system of incentives to work we would need some sort of very effective administrative structure to calculate and allocate these incentives and to use them. This might be a task beyond even the capabilities of the federal government. If we want to do this, we need to create a way to correct any irrationalities in the system. For example, if people decide that they are not going to act like rational decision-makers and are not going to use incentives, who is going to correct the system?

I stop short of calling the idea of incentives a pipe dream, but no more than I would for the idea of standards. I think we have standards precisely because we can not do the things that we need to do without them.

Russell: First, the objectives are already there. You do not need to go through the standard setting exercise again to have a system that achieves the objective in a different way. I am not suggesting that we rethink our objectives. I am suggesting that we try to obtain them in a more efficient way, a more flexible way.

Secondly, it is exactly because I think the administrative problems in the long run would be considerably less, that I advocate that we think about marketable permits. Certainly, if I believed that the use of marketable permits was going to create a bigger administrative problem than what we have now, I would never advocate it all, because I think what we have now is a nearly hopeless administrative muddle that will never get us anywhere in the long run. In the short run it has gotten us some place but we are coming up against the limits of what we can achieve. We are forcing the EPA into a position where it has to have a nearly infinite amount of information about the finest grained detail in the economy — the level of the plant and what it does and what it can do — and then expecting the EPA to make very wise judgments which can stand up in court about what ought to be done. I think that imposes loads on state and federal administrators that they should not be asked to bear.

Now, with regard to correcting for irrationality, I guess it is partly a matter of taste whether you think that is a really big problem or not. My own predilection is to think that people in general are going to act rationally and seek their own self-interest if we give them the chance. There are a lot of people, even people who worry about marketable permits in a more positive way than you do, who think that irrationality is a big problem, and they have proposed particular ways around it. So it is something that has occurred to me and to them, but it does not bother me very much.

Comment: Let me amplify what I said. I did not say that you could not devise a system where incentives would somehow do the job. What I said was, "in order to show that it was better" we would have to satisfy those three things. I do not think that can be done. I am sure we can have an incentive system, there is no problem about that, but whether or not it is a better system would have to be proven.

Russell: What would be your definition of "better"?

Comment: To which objective function does the term "better" refer? I might have one and you might have another.

Russell: Let us take a given set of ambient standards. Let us say we want to meet those standards and in three ways we want to improve on the current situation. We want to have more flexibility to deal with the growth and change that is a fact of life. We want to stimulate technological change; change in the direction of less polluting technologies, more efficient treatment processes, and that sort of thing. And we want to reduce the information and decision load on government agencies. Those improvements define my objective function. They are what I think we can do better with an incentive system.

I am no welfare-economist dreamer who thinks we can measure damages and have the perfect system where marginal damages equal marginal costs. I would never be caught in the trap of asserting that.

Comment: Perhaps in a brief statement I could explain the rules we are pursuing in Europe concerning these problems. The European Community Countries (EEC) have distinguished what we call toxic substances and very toxic substances, the last category being equal to the annex I in various conventions, and those substances should be dealt with and cleaned up at the point source.

It has been agreed upon by the member countries (with one exception) that substances, such as cadmium, mercury, many organohalogen compounds, and so on, should be cleaned at the source, and it can really be done. You can look into several industrial processes and see that this can be done.

A very broad directive on pollution caused by certain dangerous substances discharged into the aquatic environment of the EEC was adopted in 1976. Related directives, concerning specific regulation for the import and use of PCB and PCT, for example, have been passed by the Council in Bruxelles, and similar regulation has been accepted in the Paris and the Baltic conventions. Thus, in my opinion, we are dealing with certain pollution problems in a more uniform manner in Europe.

The other kinds of substances, a variety of heavy metals, cyanides, fluorides and so on, should be dealt with mainly from the standpoint of the receiving water — and water quality objectives. The EEC member countries are coming up with water quality objectives for the various uses of their coastal waters; that is, objectives for bathing water, shellfishing areas, and so on.

It is my opinion that, to a larger degree, we have a specific policy in Europe, and the Paris convention, of which many of EEC countries as well as Norway, Sweden, Spain, and Portugal are members, reflects very much what is going on inside the EEC countries. This leads me to the reflection of how independent the states act over here compared to the above-mentioned countries. From Europe one looks upon the United States as a uniform country, but it seems to me that the Commission in Bruxelles has been given more power than the U.S. federal government in some aspects, and is able to deal with at least some problems in a more uniform way. It is my opinion that when it comes to environmental questions there are some good reasons for pursuing the problems from this broad regulation view.

Knauss: I became interested in the problem of marine pollution policy a couple of years ago when it became clear to me that the very tough and often absolute prohibition standards in the present ocean legislation were going to be very difficult to maintain for a number of reasons, not the least of which was that new requirements for land disposal were going to make it necessary to reconsider all of our alternatives. What we have been hearing over the past two or three days is that this reconsideration is indeed happening, or is about to happen. The words we are now using are "balanced approach to waste management" and that kind of thing. As one makes the slight change in emphasis, one realizes that perhaps the ocean can absorb a fair amount of pollutants.

As a number of people have said here, decisions are based on value judgments, and scientific findings are only one ingredient in a value judgment made by society. Public interest groups and different lobbies play a very important role in shaping these value judgments. As we look more and more at the question of where we are going to put our waste, I believe we will look more and more at the ocean as the solution to a number of waste management problems. One of my concerns — not an immediate concern but one we might have five or eight years from now — is that there will be pressure to put almost everything in the ocean. The problem with fighting that kind of value judgment, if you are so inclined, is that the ocean constituency is very small, and probably very weak. Fish don't vote.